PROGRAMMED BEGINNING ALGEBRA

PROGRAMMED BEGINNING ALGEBRA

SECOND EDITION

UNIT I Natural Numbers

UNIT II Integers

Irving Drooyan & William Wooton

Pierce College, Los Angeles

JOHN WILEY & SONS, INC. New York • London • Sydney • Toronto

Library of Congress Catalogue Card Number: 70-152495

ISBN 0-471-22300-X

Printed in the United States of America

10 9 8 7 6 5

PREFACE

Programmed Beginning Algebra consists of eight units of material, covering topics normally a part of a course in first-year algebra. The material is organized as follows:

Volume	Unit	Topic
Volume I	Unit I	Natural Numbers
	Unit II	Integers
Volume II	Unit III	First-Degree Equations and Inequalities in One Variable
	Unit IV	Products and Factors
Volume III	Unit V	Fractions and Fractional Equations
Volume IV	Unit VI	Graphs and Linear Systems
Volume V	Unit VII	Radicals
	Unit VIII	Quadratic Equations

This second edition of *Programmed Beginning Algebra* differs from the first edition in a number of ways, although the basic program, which was developed empirically over a long period of time and which has statistical reliability, has not been altered.

Numerous exercise sets have been added and appended to each unit. These exercise sets are optional reinforcements in the performance of routine algebraic processes, and they have been deliberately kept out of the program proper so as not to interfere with the basic learning sequence. Students may or may not use these exercise sets as is appropriate in each individual case. Page margins have been shaded to help the user locate exercise sets as well as other material in the appendix of each unit.

The physical size of each volume of the second edition has been reduced for convenient use without sacrificing the usability of the type. Also, shading has been introduced in the response column to help identify clarifying remarks that are not part of the response per se.

This program can be used in a variety of ways. For reasonably capable students who accomplish the entire program, it constitutes a complete first-year course in algebra. Preferably, the program should be supplemented with periodic classroom sessions or individual conferences with a qualified mathematics instructor; but, even without such assistance, a conscientious student can expect to achieve a degree of proficiency adequate to perform in a satisfactory manner on most standard achievement tests in beginning algebra.

Any one of the eight units can be used independently of the others. Each unit can be used as a self-tutor, to supplement and/or reinforce the same topic in any standard first-year algebra textbook. If classroom attendance is interrupted due to illness or for some other reason, the material missed in class can be covered by accomplishing the appropriate unit in this series. The statement of objectives and the table of contents at the beginning of each unit will give a good picture of the material covered herein.

The material, as presented, is somewhat traditional in point of view. However, set terminology and set symbolism are used where helpful, inequalities in one and two variables are discussed in conjunction with equations, solution sets of equations and their graphs are approached through the concept of ordered pairs, and emphasis is placed on the logical basis for the routine operations performed.

The units in the program are sequential, and, with the exception of Unit I, there are certain prerequisites necessary for each. Every volume, except Volume I, includes a prerequisite test, together with a suggested minimum score. This score is to be viewed as flexible. The primary purpose of the test is to provide the individual student with a means of determining whether or not he can expect to complete the material in the volume without undue difficulty.

IRVING DROOYAN
WILLIAM WOOTON

HOW TO USE THIS MATERIAL

The theory underlying programmed instruction is quite simple. The program presents you with information in small steps called frames, and, at the same time, requires you to make frequent and active written responses in the form of a word, a phrase, or a symbol. Programmed material is only as effective as you, the student, make it. In order to achieve the maximum benefit from this unit, you should follow the instructions carefully.

If you wish to measure what you learn, you should take one form of the self-evaluation tests at the end of each unit before beginning that unit. Then, upon completion, you can take the alternate form of the test and compare the two scores.

To progress through the program:

1. Place the response shield over the response column (the column on the left-hand side of the page) so that the entire column is covered.
2. Read the first frame on the right-hand side of the page carefully, noting, as you do, the place where you are asked to respond. Do not write a response until you have read the entire frame.
3. After deciding on the proper response, write it in the blank provided or at the end of the frame.
4. Slide the response shield down until it is level with the top of the next frame. This will uncover the correct response to the preceding frame and any additional remarks accompanying it.
5. Verify that you have made the correct response. You should almost always have done so, because the frames were constructed in such fashion that it will be difficult for you to make an error.
6. If you have made the correct response, repeat Steps 2 to 5 above with the next frame. If you have made an error on the frame, read the frame again, draw a line through your incorrect response and write the correct response. Following this, repeat Steps 2 to 5 above.

Essentially, this is all there is to your part of the task, and the program must assume the remainder of the responsibility for your learning. However, there are a few additional things you can do to improve the effectiveness of the material.

1. Do not try to go too fast. Each frame should be read carefully and you should think about the response you are going to make. Does it fit the wording of the frame? Does it make sense? In many frames you will find clues (called prompts) that are designed to aid you in making a correct response. Does your proposed response match the clue?

2. Mathematics is a language written in symbols. In order to understand mathematics, you have to make a conscious effort to read the symbolism as though it were written out in words.

3. Do not try to do too much at once. If you stay with your work too long at one time, you tend to become impatient, and the number of errors you make will probably rise. If you are making good progress, and are not tired, there is no reason to stop, but if you find yourself becoming bored, or find yourself making a high percentage of errors, take a break. Come back to the program later.

4. After taking a break, pick up your work a few frames behind where you stopped work. This will lead you back into the logical sequence with a minimum of confusion.

5. Do not be overly attentive to the remarks that accompany some of the responses in the response column. They are there to clear up points of confusion once in a while, but if you are not confused, don't waste time on them. The important thing as far as you are concerned is the response itself.

6. If you wish additional reenforcement with respect to certain parts of the program, you can complete the supplemental exercises that are provided at the end of each unit. Footnotes are used at appropriate places in the program to refer you to the exercises.

7. Keep note paper near at hand to enable you to solve problems that require several steps.

8. Complete the unit. If you do not complete the unit, you cannot expect to achieve the full benefit of the material.

UNIT I Natural Numbers

OBJECTIVES

Upon completion of the unit the student should:

1. Know what is meant by “natural number,” “prime number,” and “composite number,” and be able to identify representative members of each set.

2. Understand the meaning of, and be able to apply, the
   a. Commutative law of addition
   b. Associative law of addition
   c. Commutative law of multiplication
   d. Associative law of multiplication
   e. Distributive law

3. Be able to write sums, differences, products, and quotients involving numerals and variables.

4. Be able to write monomials in both exponential and completely factored forms.

5. Be able to simplify expressions by combining like terms.

6. Be able to define and identify examples of “exponent,” “coefficient,” “base,” “power,” “monomial,” “binomial,” “trinomial,” “polynomial,” and “expression.”

7. Know the meaning of the symbols “$<$,” “$=$,” and “$>$,” and be able to read and write simple mathematical sentences involving these relationships.

8. Be able to graph natural numbers.

CONTENTS

UNIT I
Natural Numbers

Remark. This unit introduces a part of the vocabulary necessary for the successful study of algebra. The terms are introduced through a consideration of the numbers of arithmetic, and certain properties associated with these numbers.

set

1. Any collection of things is called a set. The letters in the English alphabet form a _____ .

set

2. The collection of coins in a man's pocket is a _____ .

8; 10; 12

3. Braces, { }, are used to identify a set. Thus, {1, 2, 3, 4, 5} is the set of numbers 1, 2, 3, 4, and 5. Similarly, {8, 10, 12} denotes the set of numbers _____, _____, and _____ .

member

4. Any one of a collection of things in a set is called a member of the set. Thus, 3 is a __________ of {3, 4, 5}.

elements

5. A member of a set is also called an element of the set. Thus, 3, 4, and 5 are the members or the ______________ of {3, 4, 5}.

set; members

Or set; elements.

6. The symbolism $\{3, 4, 5\}$ is read "the set whose members are 3, 4, and 5." $\{1, 9, 15\}$ is read "the _____ whose __________ are 1, 9, and 15."

members

Or elements.

7. $\{2, 4, 6, 8, 10\}$ has five members. $\{1, 7, 9\}$ has three __________.

natural

8. The set of all numbers used to count objects is called the set of natural numbers. Thus, 1, 2, 3, and 247 are examples of natural numbers. 5 is also a __________ number.

natural

9. 243 is a member of the set of __________ numbers.

is not

10. Numbers such as 1/2, 3/4, 2.17 are not used to count objects and are not natural numbers. 4/5 (is/is not) a member of the set of natural numbers.

is

11. 17 (is/is not) a natural number.

is not

11/3 is not used to count things.

12. 11/3 (is/is not) a natural number.

is not

13. 7/5 (is/is not) a member of the set of natural numbers.

is

You can count to this number if you take enough time.

14. The number 1,000,001 (is/is not) a natural number.

171

15. Circle the natural number that is a member of {1.5, 33/5, 171}.

2

16. Circle the natural number that is a member of {2/3, 2, 1.7}.

23; 21

17. The natural numbers that are members of {23, 0.1, 21, 45/2} are ____ and ____ .

finite

18. If the members of a set can be counted 1, 2, 3, etc., and if the set has a last member, it is called a <u>finite set</u>. {1, 2, 3, 4} contains four members and is a ________ set.

finite set

19. {3, 4, 5, 6, 7, 8} is a ________ ____ .

infinite

20. Sets whose elements cannot be counted, that is, which contain no last member, are called <u>infinite sets</u>. Since in counting the natural numbers one never arrives at a last number, the set of natural numbers is an ________ set.

infinite set

21. Some infinite sets of numbers (such as the set of natural numbers) can be represented by using three dots within the braces. Thus, {1, 2, 3, 4, . . . } represents the set of natural numbers, where the three dots mean that the numbers continue indefinitely and that the set does not contain a last member. {2, 4, 6, 8, 10, . . . } is an ________ ____ .

finite

22. {5, 10, 15, 20} is a ________ set that contains four members.

infinite set

23. {5, 10, 15, 20, . . . } is an ________ ____ .

Remark. Having established what is meant by the word "set," we are now ready to start using letters to represent numbers.

variable	**24.** When a letter such as a or b is used to represent any member of a given set, it is called a variable. Thus, if a is used to represent a member of the set of natural numbers, a is a ______.
variables	**25.** Any letter may be used as a variable. The letters x, y, and z are most frequently used. When used to represent an unspecified member of a set, the letters are called ______.
equals	**26.** The symbol "=" is used to represent the words "is equal to" or "equals." Thus, $3 = 3$ is read "3 is equal to 3" or "3 ______ 3."
3	**27.** If $x = 3$ is true, the variable x represents the number ____.
y	**28.** If $y = 7$ is true, the variable ____ represents the number 7.
replacement	**29.** A variable represents any element of a given set of numbers. The set of numbers is called the replacement set of the variable. Thus, if x represents a number from $\{1, 2, 3\}$, $\{1, 2, 3\}$ is the ______ set of x.
$\{1, 2, 3, 4, 5\}$	**30.** If x represents an element of the set $\{1, 2, 3, 4, 5\}$, the replacement set of x is $\{$ ______ $\}$.
replacement	**31.** If y represents a member of $\{10, 15, 20, 25\}$ then $\{10, 15, 20, 25\}$ is the ______ set of y.

Remark. Now that we have established what is meant by the word "variable," we want to take a long, hard look at the operation of addition. The addition of natural numbers is governed by a very few laws, but the laws are quite important, and the ideas they represent are employed everywhere in algebra.

addition

32. The symbol "+" is used to indicate the operation of ________________ .

+

33. The operation of addition is indicated by the symbol____.

sum

34. The number obtained by adding one number to another is called the sum of the two numbers. If 8 is added to 2, the________ is 10.

sum

35. If 15 is added to 13, the________ is 28.

six plus seven

36. The expression 5 + 8 is read "five plus eight" and denotes the sum of 5 and 8. 6 + 7 is read "_______ _______ _________."

8

37. Notice that the sum of 5 and 3 is 8 and that the sum of 3 and 5 is also 8. That is, 5 + 3 = 3 + 5. Similarly, 8 + 6 = 6 +____.

commutative

38. When the result of applying an operation on two numbers is the same regardless of the order in which the numbers are operated upon, the operation is said to be commutative. From the example 8 + 3 = 3 + 8, it appears that the addition of natural numbers is __________________ .

commutative

39. Since no case can be found where the sum of two natural numbers is different if the order in which they are added is different, it can be assumed that the addition of any two natural numbers is ________________.

axiom

40. When an assumption is made in mathematics, it is customary to call the assumption an axiom. Thus, the statement that the order in which any two natural numbers are added does not alter their sum could be called the commutative __________ of addition.

axiom

41. An axiom is not subject to proof. It is not proved, but assumed, that the sum of two natural numbers is another natural number. An assumption in mathematics is called an __________ .

assumption

42. An axiom is an ________________.

law

43. Axioms are sometimes called laws. Thus the commutative axiom of addition is sometimes called the commutative ________ of addition.

commutative

44. The commutative law of addition may be stated formally as follows:

The sum of any two natural numbers a *and* b *is the same as the sum of* b *and* a.

To say that $2 + 3 = 3 + 2$ is an application of the ________________ law.

commutative law

45. The commutative law of addition is usually stated:

If a *and* b *are natural numbers,*

$$a + b = b + a$$

where a and b represent natural numbers. To assume that $105 + 315$ is the same as $315 + 105$ is an application of the ________________________ of addition.

axiom

Or "assumption."

46. Because it is not proved that $a + b = b + a$ for all natural numbers a and b, the commutative law is an ______.

$b + a$

47. If a and b represent natural numbers, the commutative law of addition says that $a + b =$ ______.

$a + b$

$b + a$ is also correct.

48. The sum of 13 and 12 is 25. The sum of x and y cannot be represented by a numeral until values for x and y are known. However, the symbols $x + y$ can be used to represent the sum. The sum of a and b can be written ______.

9

49. In arithmetic, 2 + 7 is looked at as a problem, that of adding 2 to 7. In algebra, we must get used to looking at 2 + 7 as simply another way to write the single numeral ___.

8 + 1

50. Other ways to write 9 are 5 + 4, 6 + 3, or ___ + 1.

$r + s$

$s + r$ is also correct.

51. The sum of two natural numbers represented by r and s is written ______.

commutative

52. Notice that the sum of two natural numbers represented by r and s can be written either $r + s$ or $s + r$. This is an application of the ______ law.

$u + v$

Or $v + u$.

53. The sum of the natural numbers u and v is written ______.

added	**54.** The operation of addition applies only to two numbers at a time. To find a single numeral representing 3 + 4 + 5, the first two numbers can be added to give 7, and then 7 can be ________ to 5 to give 12.
adding	**55.** A single numeral, 12, representing 3 + 4 + 5 can be found by first adding 4 and 5 to obtain 9 and then ________ 9 to 3.
two	**56.** To add any given collection of numbers, the operation of addition is performed on only _____ numbers at a time.
first	**57.** To show that two numbers are to be considered as grouped for the purpose of addition, the symbols ( ), called <u>parentheses</u>, are used. Thus (5 + 4) + 7 means that 5 and 4 are to be added first. 5 + (4 + 7) means that 4 and 7 are to be added ________.
parentheses	**58.** The symbols ( ) are called ________________.
parentheses; brackets	**59.** Parentheses are examples of grouping devices. Other symbols used for this purpose are <u>brackets</u>, [ ]. For example, [(3 + 4) + 8] + 5 means that 3 is to be added to 4, the result, 7, is to be added to 8, and the result of this, 15, is to be added to 5. The symbols ( ) are called ________________ and the symbols [ ] are called ____________.
17	**60.** To add three numbers 8 + 3 + 6, the numbers can be grouped (8 + 3) + 6 or 8 + (3 + 6). The result in both cases is the number _____.

associative

61. It is assumed that for any three natural numbers x, y, and z,

$$(x + y) + z = x + (y + z)$$

This assumption is called the associative law of addition. The fact that 8 + 2 + 4 can be viewed as either (8 + 2) + 4 or as 8 + (2 + 4) is an application of the ________________ law of addition.

associative

62. The fact that 8 + 10 + 3 can be viewed as 18 + 3 or as 8 + 13 is an application of the ________________ law of addition.

commutative; associative

63. Two basic laws apply to the addition of natural numbers. 2 + 3 = 3 + 2 illustrates the ____________ law, and (2 + 3) + 4 = 2 + (3 + 4) illustrates the ____________ law.

commutative

64. $r + s = s + r$ is a statement of the ____________ law of addition.

associative

65. The fact that $r + s + t$ means either $(r + s) + t$ or $r + (s + t)$ is a statement of the ____________ law of addition.

order

66. The commutative law of addition is concerned with the order of the terms. Thus, when $x + 2$ is written $2 + x$, only the ________ of the terms has been changed.

grouping

67. The associative law of addition is concerned with the grouping of the terms. Thus, when $(x + y) + 2$ is written $x + (y + 2)$, only the __________ of the terms has been changed.

commutative

Only the <u>order</u> is changed.

68. The law that permits writing $3 + y$ as $y + 3$ is the________________ law of addition.

associative

Only the <u>grouping</u> has been changed.

69. The law that permits writing $(2 + x) + y$ as $2 + (x + y)$ is the ________________ law of addition.

commutative

Do you see that only the <u>order</u> has been changed, not the <u>grouping</u>?

70. $(x + y) + z$ may be written as $z + (x + y)$ by the ________________ law of addition.

associative

Do you see that only the <u>grouping</u> has been changed, not the <u>order</u>?

71. When $(x + y) + z$ is written $x + (y + z)$, the ________________ law of addition has been applied.

Remark. The operation of addition with natural numbers is governed by two axioms, the commutative law, and the associative law. Remember these names and what they mean, because they will be used over and over again in this and later units. Next, we shall examine two properties of the equality relation that are also very useful.

number

72. A statement of equality means that the expressions that are set equal are simply different names for the same thing. Thus, $5 + 2 = 7$ means that $5 + 2$ and 7 are names for the same n________r.

$8 + 3$

73. Since $5 + 2$ and 7 are different names for the same number, $5 + 2 = 7$ can also be written $7 = 5 + 2$. It makes no difference which expression appears on the right and which on the left of the equals sign. Thus, $8 + 3 = 11$ can be written $11 =$ ________.

$5 + 3$

74. $8 = 5 + 3$ can also be written ________ $= 8$.

$x + y$

75. $x + y = z$ can be written $z =$ ________.

b

76. $a = b$ can be written ____ $= a$.

symmetric

77. The fact that if $a = b$, then $b = a$, is important enough to be given a name. It is called the symmetric law of equality. Thus, if $8 + 5 = 13$ is written $13 = 8 + 5$, the ________________ law of equality has been applied.

symmetric

78. Writing $8 - 4 = 4$ as $4 = 8 - 4$ is an application of the ________________ law of equality.

$s = r$

79. By the symmetric law of equality, if $r = s$, then ____ = ____.

$a + b + c = x$

80. If $x = a + b + c$, then by the symmetric law of equality, ________________ = ______.

axioms

Did you remember?

81. The symmetric law of equality is an assumption; that is, we assume it is always true. In mathematics, assumptions such as this are called ______________.

substitution

82. Another important assumption is called the substitution axiom. This axiom asserts that if $x = y$, then either may be replaced by the other in any mathematical expression without changing the truth or falsity of the expression. Thus, if $x = y$, the expression $x = 6$ may be written as $y = 6$, where x has been replaced by y by an application of the ________________ axiom.

x

83. The substitution axiom states that if $a = b$ and $b = c$, then $a = c$. Similarly, if $x = y$ and $y = 4$, then ___ $= 4$.

20

84. If $a = c$ and $a = 20$, then $c =$ ____.

substitution

85. If $r = s$ and $s = 2$, then $r = 2$. This is an application of the ____________ axiom.

replace

Or substitute

86. There is a close relationship between "substitution" and "replacement." Thus, when 3 is substituted for x in $x + 2 = 5$, we simply ________ x with 3 and obtain $3 + 2 = 5$.

$4 + a$

87. If $x = y$ and $y = 4 + a$, then $x =$ ______.

5

88. If $x = y$ and $x + 2 = 5$, then $y + 2 =$ ____.

substitution

89. If $y = 7$ and $y = x$, then by the ________ axiom, $x = 7$.

Remark. We are now going to take a very brief look at the operation of subtraction, just long enough to establish what is meant by "subtract." Following this, we shall make a rather lengthy study of the operation of multiplication.

difference

90. The symbol "–" is used to denote the operation of subtraction. The result of subtracting one natural number from another is called the difference of the numbers. Thus, $5 - 2, = 3$ (read "five minus two equals three") means that the ____________ of 5 and 2 is 3.

difference

91. If 4 is subtracted from 9, the ________ is 5.

3

92. In algebra, it is desirable to view the symbols $a - b$ as representing a number, the difference of a and b. Thus, $5 - 2$ can be looked upon as representing the same number represented by the single numeral ____.

$c - d$

93. If c and d represent natural numbers, the result of subtracting d from c is written ________ .

$\times$

94. The symbol "$\times$" is used to denote the operation of multiplication. To show that 3 is to be multiplied by 5, the symbols 3 ____ 5 are used.

six times seven

95. The expression 3×5 is read "three times five." 6×7 is read "________________."

product

96. The result of multiplying one number by another is called the product of the two numbers. Thus, 15 is the ____________ of 3 and 5.

factors

97. The numbers multiplied together to yield a product are called factors of the product. 7 and 3 are factors of 21 because $7 \times 3 = 21$. Because $6 \times 7 = 42$, 6 and 7 are ____________ of 42.

product

98. In arithmetic, the symbols 4×5 denote the operation of multiplication. In algebra, it is also convenient to think of 4×5 as a product, simply a different way to represent the number 20. Thus, 6×7 is a ____________ .

factors

99. 9×7 represents a product. 9 and 7 are________ of the product.

8; 4

100. The factors of the product 8×4 are____ and____.

times

101. To indicate multiplication in algebra, the multiplication sign is rarely used. Instead, the product of two factors, at least one of which is a variable, is usually indicated by simply writing the factors side by side. $3y$ means "three times y," ab means "a __________ b."

product

102. xy represents the product of x and y; rs represents the______________ of r and s.

times

103. Multiplication is not intended when numerals are written side by side. Thus 23 means the number twenty-three and not "2 times 3." To indicate the multiplication of 2 and 3, the expression $2 \cdot 3$ may be used. $2 \cdot 3$ means "2__________ 3."

product

104. $5 \cdot 7$ represents the ______________ of 5 and 7.

times

105. Another means of expressing multiplication involves the use of parentheses, ( ). For instance, 5(7) means "5 times 7." $6 \cdot 8$ and 6(8) both mean "6 __________ 8."

factors

106. $9 \cdot 10$ and 9(10) both indicate the multiplication of the f ____________ 9 and 10.

11; 11

107. Sometimes both factors are included in parentheses. (3)(12) means the same as 3(12) or $3 \cdot 12$. For another example, $5 \cdot 11$ can be written as 5(____) or (5)(____).

3

108. Multiplication is commutative; that is, the order of multiplying two numbers does not affect the product. Thus, $(3)(7) = (7)(\_\_)$.

commutative

109. The commutative law of multiplication asserts:

If a, b, *and* c *are natural numbers*,

$$ab = ba.$$

$5 \cdot 3 = 3 \cdot 5$ is an illustration of the ____________ law of multiplication.

commutative law

110. That $(2)(5)$ and $(5)(2)$ represent the same product is guaranteed by the ____________ ______ of multiplication.

$3a$

111. When an expression for a product contains both a numeral and a variable, it is preferable to write the numeral first. Although the product of 3 and a can be denoted by either $3a$ or $a3$, the preferred form is ____.

$6y$

112. The preferred form for representing the product of y and 6 is ____.

rs

r and s are in alphabetical order.

113. When a product contains two variables, it is customary to write the symbols in alphabetical order. Thus, while rs and sr both represent the product of r and s it is customary to write the product as ____.

xz

114. The product of z and x is written ____.

cd

115. The product of d and c is written ____.

product

116. bc represents the ____________ of b and c.

factors

117. b and c are __________ of the product bc.

product

118. rs represents a __________.

factors

119. r and s are __________ of the product rs.

associative

120. Multiplication is associative. In symbols, the <u>associative</u> <u>law</u> <u>of</u> <u>multiplication</u> is written,

$$(a \cdot b) \cdot c = a \cdot (b \cdot c)$$

where a, b, and c are natural numbers. The fact that $(2 \cdot 3) \cdot 4$ and $2 \cdot (3 \cdot 4)$ represent the same product is an example of an application of the __________ law of multiplication.

associative law

121. The fact that $3 \cdot (5 \cdot 7)$ and $(3 \cdot 5) \cdot 7$ represent the same product is an example of an application of the __________ __________ of multiplication.

order

122. Both the commutative law of addition and the commutative law of multiplication have to do with <u>order</u>. Thus, when (3)(7) is written (7)(3), only the __________ of the factors has been changed.

grouping

123. Both the associative law of addition and the associative law of multiplication have to do with <u>grouping</u>. Thus, when $(3 \cdot 7) \cdot 2$ is written $3 \cdot (7 \cdot 2)$, only the __________ of the factors has been changed.

commutative; multiplication

124. The law that permits writing (7)(8) as (8)(7) is the __________ law of __________.

commutative; addition

125. The law that permits writing $8 + 6$ as $6 + 8$ is the __________ law of __________.

associative;
multiplication

126. The law that permits writing $(8 \cdot 6) \cdot 9$ as $8 \cdot (6 \cdot 9)$ is the ______________ law of ______________.

associative;
addition

127. The law that permits writing $(8 + 6) + 9$ as $8 + (6 + 9)$ is the ______________ law of ______________.

$2(3 \cdot 9)$

128. When a numeral or variable is written next to a set of grouping symbols, ( ) or [ ], the operation of multiplication is understood. Thus, $5 \cdot (6 \cdot 7)$ can be written as $5(6 \cdot 7)$. $2 \cdot (3 \cdot 9)$ can be written as ___$(3 \cdot 9)$.

$3(4 \cdot 5)$

129. $3 \cdot (4 \cdot 5)$ can be written as ___$(4 \cdot 5)$.

product

130. 3(5) represents the __________ of 3 and 5.

$(2 \cdot 3)9$

$(2 \cdot 3) \cdot 9$ is also correct.

131. By the associative law of multiplication, the factors $2 \cdot 3 \cdot 9$ can be grouped as $2(3 \cdot 9)$ or as __________.

4

132. $4rs$ means either $(4r)s$ or ___ (rs).

single

133. Grouping devices are used to group numbers when the group is to be viewed as a single number. $(x + y)$ indicates that the sum of the numbers x and y is to be thought of as a single number. $(x + 3)$ means that the sum of x and 3 is to be thought of as a __________ number.

multiplied

134. The product of 3 and the sum of x and 2 can be written $3(x + 2)$. Because there is no sign of operation between 3 and the parentheses, the symbols mean that 3 is to be ______________ by $(x + 2)$.

multiplied

135. $4(y + 3)$ means that 4 is to be ________ by $(y + 3)$.

product

136. $2x(a + b)$ represents the ________ of $2x$ and $(a + b)$.

$7(2x + y)$

137. Write the product of 7 and $2x + y$.

$a(x + 3)$

138. Write the product of a and $x + 3$.

Remark. We have seen that there are commutative laws and associative laws for both addition and multiplication in the set of natural numbers. There is one more law that is fundamental to both operations because it establishes a relationship between them.

8

139. The product $3(1 + 4)$ can be computed by first adding 1 and 4 to obtain 5 and then multiplying 3 by 5 to obtain 15. The product $2(3 + 1)$ can be written $2(4)$, which is equal to___.

16

140. $2(3 + 5) = 2(8) =$_____. $2(3 + 5)$ can also be computed by multiplying 2 times 3 and 2 times 5 and adding the resulting products. Thus, $2(3 + 5) = 2(3) + 2(5) = 6 + 10 =$_____.

1; 4

141. $2(3 + 5) = 2(3) + 2(5)$. Similarly, $5(1 + 4) = 5(\_\_) + 5(\_\_)$.

4; 4

142. $4(8 + 2) = \_\_(8) + \_\_(2)$.

7; 5

143. $3(7 + 5) = 3(\_\_) + 3(\_\_)$.

distributive

144. Because this relationship between sums and products is so important in mathematics, it is taken as an axiom and is called the <u>distributive law</u>. The distributive law asserts that:

If a, b, *and* c *are natural numbers*,
$$a(b + c) = ab + ac.$$

Thus, the statement $5(1 + 4) = 5(1) + 5(4)$ is an application of the ________________ law.

distributive

145. $3(7 + 5) = 3(7) + 3(5)$ is an application of the ________________ law.

2; 2

146. $2(6 + 3) =$ ___ $(6) +$ ___ (3).

3; 5

147. By the distributive law, if x represents a natural number, $x(3 + 5) = x($___$) + x($___$)$.

3; 3

148. $3(x + y) =$ ___ $(x) +$ ___ (y).

distributive

149. $4(x + y) = 4x + 4y$ is an application of the ________________ law.

Remark. This brings us to the last of the basic arithmetic operations, division.

quotient

150. The result of dividing one number by another is called the <u>quotient</u> of the numbers. When 8 is divided by 4, the ________________ is 3.

quotient

151. If 12 is divided by 4, the ________________ is 3.

divided by

152. The symbol "÷" is used to indicate division. $6 \div 3$ means "6 divided by 3." $24 \div 6$ means "24 ______________ ________ 6."

18; 3

153. $6\overline{)24}$ and $\frac{24}{6}$ are other ways of indicating division. $3\overline{)18}$ means "______ divided by____."

24; 6

154. The <u>fraction</u> is the most useful symbolism for division. $\frac{24}{6}$ is generally more convenient to use than $24 \div 6$ or $6\overline{)24}$. $\frac{24}{6}$ means " ______ divided by_____."

$\frac{5}{3}$

155. "5 divided by 3" would be indicated by the fraction______ .

$\frac{x}{y}$

156. "x divided by y" would be indicated by the fraction______ .

is not

157. Multiplication and addition are commutative; that is, $xy = yx$ and $x + y = y + x$. Division, however, is not commutative. Thus $12 \div 4$ (is/is not) the same as $4 \div 12$.

is not

158. $\frac{4}{3}$ (is/is not) the same as $\frac{3}{4}$.

quotient

159. $\frac{x}{y}$ can be looked at either as a direction to divide x by y or as representing the quotient of x divided by y. $\frac{z}{3}$ represents the______________of z divided by 3.

$\frac{7}{a}$

160. The quotient of 7 divided by a is written ______.

quotient

161. $\frac{3r}{2}$ represents the ____________ when the product $3r$ is divided by 2.

product

162. $\frac{xy}{z}$ represents the quotient when the __________ xy is divided by z.

$x + 2$

163. $(x + 2) \div 3$ means that __________ is to be divided by 3.

quotient; $a + b$

164. $7 \div (a + b)$ represents the ______________ when 7 is divided by __________ .

$x + 3$

165. If a quotient is written as a fraction, such as $\frac{x + 3}{7}$, parentheses are not needed because the fraction bar serves the same purpose. $\frac{x + 3}{7}$ means that __________ is to be divided by 7.

$a + b$; 3

166. $\frac{a + b}{3}$ represents the quotient when __________ is divided by____.

$\frac{x + y}{z}$

167. Write the quotient of $x + y$ divided by z in fractional form.

$\frac{5}{2x + 1}$

168. Write the quotient of 5 divided by $2x + 1$.

$\frac{2x+1}{4y-2}$

169. Write the quotient of $2x+1$ divided by $4y-2$.

Remark. Now that we are familiar with the vocabulary and the basic laws associated with the operations of arithmetic, our next task is to begin relating these operations to variables, and to combinations of variables and numerals. First we will look at certain kinds of natural numbers and special ways to represent them.

exactly

170. To say that one natural number is exactly divisible by another means that the division does not involve a remainder. Thus, 8 is___________ divisible by 2, because there is no remainder.

1

171. When 7 is divided by 2, the remainder is____.

1

172. The only natural numbers that divide into 7 without leaving a remainder are 7 and____.

prime

173. A natural number greater than 1 that is exactly divisible by itself and 1 only is called a <u>prime number</u>. Therefore, 2, 3, 5, 7, and 11 are_____________ numbers.

2

174. The number 1 is not considered a prime number or a composite number for reasons of importance in more advanced work. Therefore, the smallest prime number is____.

prime; 2

175. 6 is not a prime number because it is divisible by 2 and 3 as well as by itself and 1. 8 is not a _____________ number because it is divisible by 4 and_____ as well as itself and 1.

5; 2

Or 2; 5

176. 10 is not a prime number because it is divisible by___ and___ as well as by itself and 1.

is not

14 is divisible by 2 and 7.

177. 14 (is/is not) a prime number.

7; 13

178. Circle the prime numbers in the set {4, 7, 13, 15}.

17; 23

179. Circle the prime numbers in the set {8, 15, 17, 23}.

composite

180. A natural number greater than 1 that is not a prime number is called a composite number. A composite number, therefore, must be exactly divisible by some natural number other than itself and 1. 6 is a________________ number because it is divisible by 2 and 3 as well as itself and 1.

prime; composite

181. Any natural number greater than 1 is either a prime number or a composite number. The numbers 2, 7, and 17 are___________numbers, while 4, 8, and 9 are________________ numbers.

factors; product

182. Recall that when two or more numbers are multiplied, the result is called a product, while the numbers being multiplied are called factors. Since (3)(5) = 15, 3 and 5 are_________ of the _________ 15.

factors

183. 5 and 7 are__________of the product 35.

7

184. Two factors of 14 are 2 and____.

factors

185. 7 and 6 are __________ of the product 42.

product

186. 2 and 5 are factors of the ______________ 10.

factored

187. (2)(5) is called a factored form of the product 10. (3)(5) is a _________________ form of 15.

factored form

188. (3)(11) is a ______________ __________ of 33.

factors

189. 4 and 3 are factors of 12. 2, 2, and 3 are also __________ of 12.

prime

190. 2, 2, and 3 are called prime factors of 12 because they are themselves prime numbers. 2, 3, and 3 are ___________ factors of 18.

21

(3)(7) = 21

191. 3 and 7 are prime factors of _______.

prime factors

192. 5 and 7 are ______________ _________ of 35.

5

193. The prime factors of 30 are 2, 3, and _____.

2; 7

Or 7; 2.

194. The prime factors of 28 are 2, ___, and _____.

2; 2; 2; 2

195. The prime factors of 16 are ___, ___, ___, and ___.

completely

196. The factored form of a number that contains only prime factors is called the completely factored form of the number. Thus, $2 \cdot 2 \cdot 3$ is the ____________ factored form of 12.

completely factored

197. $2 \cdot 2 \cdot 3 \cdot 3$ is the ______________ __________ form of 36.

30

198. Except for the order of the factors, there is only one completely factored form of a natural number. Thus, except for order, $2 \cdot 3 \cdot 5$ is the only completely factored form of______.

(2)(2)(2)

199. The completely factored form of 8 is_________.

four

200. Repeated factors in a product such as (2)(2)(2) can be written in the more compact form 2^3, where it is understood that the factor 2 occurs three times. Similarly, (2)(2)(2)(2) can be written 2^4 and means that the factor 2 occurs _______times in the product.

(2)(2)(2)(2)

Or 16.

201. 2^3 represents the product (2)(2)(2) or 8, and 2^4 represents the product______________.

completely factored

202. (2)(2)(2) is the completely factored form of 2^3 or 8, and (2)(2)(2)(2) is the ______________ __________ form of 16.

(3)(3)(3)

Or 27.

203. 3^2 represents the product (3)(3) or 9, and 3^3 represents the product__________.

completely factored

204. (3)(3)(3) is the______________ _________ form of 3^3 or 27.

(3)(3)(3)(3)(3)

Or 243.

205. 3^5 represents the product ______________.

72

206. $2^3 \cdot 3^2$ represents the product (2)(2)(2)(3)(3) or_______.

yyy

207. x^2 represents the product xx, and y^3 represents the product______.

$bbbbb$

208. The completely factored form of a^4 is $aaaa$, and the completely factored form of b^5 is________.

$xxxyy$

209. The completely factored form of a^2b^3 is $aabbb$, and the completely factored form of x^3y^2 is_________.

(2)(3)$xyyy$

210. To completely factor $6xy^3$, the number 6 is first factored as $2 \cdot 3$, and then xy^3 is factored as $xyyy$. The completely factored form of $6xy^3$ then is ______________.

(2)(2)(3)$aaabbc$

211. The completely factored form of $12a^3b^2c$ is _________________.

(3)(3)$xxxyzz$

212. The completely factored form of $9x^3yz^2$ is _________________.

exponent

213. The numeral written to the right and a little above a second symbol to indicate the number of times the number represented by the second symbol occurs as a factor in a product is called an exponent. Thus, in the expression 3^2, 2 is an exponent. In the expression 4^3, 3 is an _______________.

5

214. In the expression 2^5, the exponent is____.

base

215. The number 2 in 2^5 to which the exponent is attached is called the base. In the expression 5^3, 3 is the exponent and 5 is the__________.

2; 4

216. The exponent of 4^2 is___and the base is___.

power

217. The exponent of 5^4 is 4 and the base is 5. The product itself is called a power and this example is read "five to the fourth power." 3^4 is read "three to the fourth____________."

power

218. x^3 is read "x to the third______________or x cubed."

base; exponent

219. 7^3 is read "seven to the third power or seven cubed"; 7 is called the__________and 3 is called the ________________of the power 7^3.

fourth power; base

220. y^4 is read "y to the____________ ____________." y is the__________and 4 is the exponent of the power y^4.

squared

221. 5^2 is read "five squared." x^2 is read "x ______________."

third; fourth power

222. c^2 is read "c squared," c^3 is read "c cubed" or "c to the___________power," and c^4 is read "c to the____________ ____________."

cubed

223. x^3 is read "x __________" or "x to the third power."

one

224. A variable such as x with no exponent indicated is understood to have the exponent 1. In the product xy^2, x occurs as a factor________ time.

one

225. In the product x^2y^3z, the factor z occurs______ time(s).

cubed

226. The product x^2y^3z is read "x squared times y__________ times z."

fourth power

227. The product $2ab^2c^4$ is read "two times a times b squared times c to the__________ ____________."

exponential

228. The factored form of the product bbb appears in exponential form as b^3. $aaaa$ in__________________ form is written a^4.

exponential

229. The__________________form of $cccc$ is c^4.

yx^2

230. Express yxx in exponential form.

a^2b^2c

231. Express $aabbc$ in exponential form.

$3ab^3c$

232. Express $3abbbc$ in exponential form.

$5xy^3z^2$

233. Express $5xyyyzz$ in exponential form.

completely factored

234. $5xyyyzz$ is called the _______________________ ________________ form of $5xy^3z^2$.

completely; exponential

235. 81 can be expressed as (3)(3)(3)(3) or 3^4. (3)(3)(3)(3) is called the ______________ factored form of the product, and 3^4 is called the ______________ form of the product. *

expression

236. Any meaningful collection of numbers, variables, and signs of operation such as $2xy + y^2$ is called an expression. Thus, $2x^2 + x + 3$ is an ______________.

expression

237. $\frac{xy}{2} - 5$ is an ______________.

terms

238. In an expression of the form $A + B + C + \ldots$, $A, B, C, \ldots$ are called terms. Thus, in the expression $2x + y + z$, $2x$, y, and z are __________ of the expression.

terms

239. $3x$ and $4y^2$ are __________ of the expression $3x + 4y^2$.

$4y^2$

240. The terms in the expression $2y^3 + 4y^2$ are $2y^3$ and ______.

four

241. The expression $x + 2y - z$ contains three terms. The expression $8x^3 - 2x^2 + x - 1$ contains __________ terms.

expression; $6x^3$; $6xy^2$

242. $3x + 6x^3 + 6xy^2$ is an ______________, and its terms are $3x$, _____, and _____.

polynomial

243. An expression made up of terms such as $3x^2$, $4x^5$, $27x$, and 3, where each term involves only the multiplication of numerals and variables, is called a polynomial. $3x + 2$ is a ______________.

* See Exercise 1, page 51, for additional practice.

is not

244. $\frac{3}{x} + 4$ is not a polynomial because it involves the operation of division with a variable. $x + \frac{1}{y}$ (is/is not) a polynomial.

is

245. $3x^2 + 2x + 5$ (is/is not) a polynomial.

is

246. An expression containing one term can be a polynomial. y^2 (is/is not) a polynomial.

is not

247. $\frac{2}{x}$ (is/is not) a polynomial.

polynomial

248. All polynomials are expressions, but not all expressions are polynomials. Thus, $\frac{x^2 + 2x}{y} + 3$ is an expression, but, because it involves division by a variable, it is not a ________________.

two; one

249. The polynomial $2x^3 + 3x + 4$ contains three terms. The polynomial $x + 7$ contains_______ term(s), and the polynomial $2x^2y$ contains_____ term(s).

monomial

250. Any polynomial containing just one term is called a monomial. Thus, $2x^2y$ is a monomial. y^2 is a_______________.

monomial

"polynomial" is also correct.

251. $3x^2$ is a_______________.

binomial

252. A polynomial containing two terms is called a binomial. $x + 3$ is a binomial. $x^2 + 2x$ is a _______________.

binomial

"polynomial" is also correct.

253. $3x^2y + 2y$ is a ______________.

monomial

"polynomial" is also correct.

254. $3x^2$ is a ____________________.

binomial

255. $3x^2$ is a monomial and $3x^2 + y$ is a __________ . Both are polynomials.

binomial; monomial

256. $x + 1$ is a ____________ and $4x^3y^2z$ is a ________________. Both are polynomials.

trinomial

257. A polynomial containing three terms is called a trinomial. $2x^2 + x + 1$ is a ______________.

binomial; trinomial

258. $x^2 + 2y$ is a ______________ and $x + 2y + z$ is a ___________________. Both are polynomials.

trinomial; monomial

259. $2a + b + c$ is a _________________ and $3x^3y^2$ is a _________________. Both are polynomials.

polynomials

260. Monomials, binomials, and trinomials are all ___________________.

coefficient

261. Any collection of factors in a term is called the coefficient of the remaining factors in the term. Thus in the term $3xy$, 3 is the coefficient of xy, x is the coefficient of $3y$, y is the coefficient of $3x$, and $3x$ is the ___________________ of y.

coefficient; a

262. In the term ab, a is the ______________ of b and b is the coefficient of____.

x

263. In the term xy^3z, the coefficient of y^3z is_____.

2

264. In expressions such as 3xy, the numerical part of the term, 3, is referred to as the <u>numerical coefficient</u>. The numerical coefficient of $2x$ is_____.

5

265. The numerical coefficient of $5x^2y$ is_____.

numerical

266. 4 is the______________coefficient of $4ab^2$.

1

267. In a term such as xy, where there is no written numeral, the numerical coefficient is understood to be 1. The numerical coefficient of a^2b is_____.

coefficient

268. The numerical______________of x^2y is 1.

2; 1

269. $2x + y^3$ is a binomial. The numerical coefficient of the first term, $2x$, is____. The numerical coefficient of the second term is___.

coefficient; exponent; x

270. $3x^5$ is a monomial. 3 is the numerical ____________ and 5 is the______________ on the base___.

1; 3

271. $x^3 + 2x^2 + 1$ is a trinomial. The numerical coefficient of the first term, x^3, is___, and the exponent is___.

are

272. Two terms that are identical in their variable factors are called like terms. Thus $2x$ and $3x$ have the same variable factor, x, and hence (are/are not) like terms.

like

273. $3xy$ and $7xy$ are__________terms.

like

274. $2x$ and $3y$ are not like terms because x and y are not the same variables. $3x^2$ and $3y^2$ are not __________terms.

are

275. $3x^2y$ and $4x^2y$ (are/are not) like terms.

The variable factors, x^2y, are the same in both terms.

like terms

276. $2x$ and $2x^2$ are not ________ ________ because x and x^2 are not identical.

$3x$; $5x$

277. The two like terms of the trinomial $3x + 4y + 5x$ are____ and____ .

$5xy$; $3xy$

278. The like terms of the polynomial $5xy + x + 3xy + y$ are____ and____ .

coefficients

Or "factors."

279. Like terms are exactly the same in their variable factors. Like terms may differ in their numerical________________ .

Remark. Expressions frequently contain like terms. Whenever this is the case, the expression can be rewritten in a briefer form.

2

280. Recall that by the distributive law $(3 + 5)2 = 3(2) + 5(2)$. Then, by the symmetric law, the members of the equation can be interchanged and the resulting equation is $3(2) + 5(2) = (3 + 5)(\_\_)$.

4

281. $7(3) + 2(3) = (7 + 2)3$. Similarly, $7(4) + 2(4) = (7 + 2)(\_\_)$.

3 + 2

282. $3(6) + 2(6) = (\_\_ + \_\_)(6)$.

10

283. $4(10) + 5(10) = (4 + 5)(\_\_)$.

7; 2

284. $7(5) + 2(5) = (\_\_ + \_\_)(5)$.

x

285. If x represents a natural number, the distributive law guarantees that $3x + 5x = (3 + 5)x$ and $2x + 7x = (2 + 7)\_\_$.

8; 3

286. If x represents a natural number, then by the distributive law, $8x + 3x = (\_\_ + \_\_)x$.

distributive law

287. If y represents a natural number, the statement $4y + 7y = (4 + 7)y$ or $11y$ is an application of the ________ ________.

7

288. $2y + 5y = (2 + 5)y$ or $\_\_\, y$.

y; $12y$

289. $3y + 9y = (3 + 9)\_\_$ or ____.

$20x$	290. $6x + 13x = 19x$ and $6x + 14x =$ ______.
$5y$	291. The term y can be considered to have a numerical coefficient of 1. Therefore, $4y + y = (4 + 1)y$ or______.
$8y$	292. $y + 6y = 7y$ and $y + 7y =$_____.
$34x$	293. $28x + 6x =$______.
$36z$	294. $20z + 16z =$_____.
$20c$; distributive	295. The statement that $16c + 4c =$______is an application of the________________ law.
$4x^5$	296. $x^5 + 3x^5 =$________.
$10z^9$	297. $2z^9 + 8z^9 =$_______.
$18ab$	298. The distributive law guarantees that $4xy + 7xy = 11xy$ and $3xy + 7xy = 10xy$. Similarly, $8ab + 10ab =$ ________.
26	299. $14abc + 12abc =$______ abc.
$10xyz$	300. $5xyz + 5xyz =$ __________.

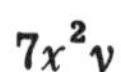

$7x^2y$

301. $3x^2y + 4x^2y =$ _________ .

14

302. The distributive law is also applicable to expressions such as $6x + 3x + 5x$; that is, it applies to any number of terms. Thus, $6x + 3x + 5x = (6 + 3 + 5)x =$ _____ x.

$14y$

303. $3y + 4y + 7y =$ _______ .

$22ab$

304. $3ab + 5ab + 14ab =$ _________ .

$11x^2y$

305. $7x^2y + 3x^2y + x^2y =$ _________ .

like

306. Recall that $6a$ and $5a$ are like terms. $7x^2$ and $2x^2$ are ________ terms.

$5a$

307. $2a$ and $3a$ are like terms. The distributive law permits writing the sum of these two products as the monomial ____.

combining; adding

Or adding; combining.

308. The terms of the expression $4a + 5a$ can be combined and written as the single product $9a$. The application of the distributive law in this form is commonly referred to as combining or adding like terms. $6a + 2a = 8a$ is an example of ____________ or ____________ like terms.

adding

309. Writing $4x + 2x = 6x$ is an example of combining or ____________ like terms.

combining

310. Writing $y^2 + 17y^2 = 18y^2$ is an example of adding or ________________ like terms.

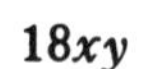

$18xy$ — **311.** Add the like terms $6xy + 12xy$.

$25x^2$ — **312.** Combine the like terms $x^2 + 15x^2 + 9x^2$.

5 — **313.** In the expression $2x + 3x + 4y$, $2x$ and $3x$ are like terms, while $4y$ is unlike either $2x$ or $3x$. Adding like terms in the expression $2x + 3x + 4y$ yields____ $x + 4y$.

11 — **314.** Like terms may be added or combined, while the sum of unlike terms can only be indicated. Thus, $4x + 7x + 2y =$_____ $x + 2y$.

$3a^2 + b^2$ — **315.** Combine like terms in the expression $2a^2 + a^2 + b^2$.

simplified — **316.** Combining or adding like terms in an expression is also commonly referred to as <u>simplifying</u> the expression. When $2x + 3y + 7y$ is written $2x + 10y$, the first expression has been__________________.

$6x + 8y$ — **317.** Simplify: $6x + 3y + 5y$.

$8a + 3b$ — **318.** Simplify: $6a + 2a + 3b$.

$7x + 5y$ — **319.** $3x + 4x + 2y + 3y =$____________.

$5a + 3b$ — **320.** $3a + b + 3a + 2b =$____________.

$9x + 3y$ — **321.** $6x + 2y + 3x + y =$____________.

$6x + 7$

322. $x + 2 + 5x + 5 =$ ________ .

$3x + 2y$

323. Parentheses are sometimes used in conjunction with sums. Thus, $(2x + y) + (x + y)$ denotes the sum of the two expressions appearing within the parentheses. This sum can be simplified by combining the like terms involved. For example, $(2x + y) + (x + y) = 2x + y + x + y =$ ________ .

$5a + 3b$

324. The sum $(3a + 2b) + (2a + b) = 3a + 2b + 2a + b =$ ________ .

$3x^2 + 5x + 3$

325. The sum $(x^2 + 2x + 1) + (2x^2 + 3x + 2) =$ $x^2 + 2x + 1 + 2x^2 + 3x + 2 =$ ________ .

$4x + 4y + 4z$

326. The sum $(x + 2y + z) + (3x + 2y + 3z) =$ ________ .

$2a^2 + 4a + 3$

327. Simplify: $(a + a^2 + 2) + (a^2 + 3a + 1)$.

$4y^2 + y + 2$

328. Simplify: $(2y^2 + 2) + (y + 2y^2)$.

$3x^2 + 4x + 2$

329. Combine like terms: $(3x^2 + 2x + 1) + (2x + 1)$.

$2x^2 + 4x + 6$

330. Combine like terms: $(x^2 + 2x + 3) + (3 + 2x + x^2)$.

$3y^2 + 3y + 6$

331. $(2y^2 + 3y + 2) + (y^2 + 4) =$ ________ .*

* See Exercise 2, page 51, for additional practice.

Remark. This unit is chiefly concerned with the properties of natural numbers. Thus far, we have studied the operations of addition, subtraction, multiplication, and division of natural numbers, and we have performed a few simple operations with polynomials over the natural numbers. As a final consideration in this unit, we shall study the order relations of natural numbers and how to associate natural numbers with points on a line.

less

332. {1, 2, 3, 4, . . . }, is an <u>ordered</u> <u>set</u>; that is, it is always possible to say that one natural number is greater than, equal to, or less than another natural number. For example, 2 is less than 5; 7 is __________ than 12.

equal

333. $\frac{6}{2}$ is__________to 3.

greater

334. 6 is__________ than 4.

less

335. The symbol "<" is used to represent the words "is less than." 4 < 5 means "4 is less than 5." 7 < 15 means "7 is__________than 15."

less than

336. 3 < 5 means "3 is ______ ________5."

equal

337. The symbol "=" is used to represent the words "is equal to." 3 = 3 is read "3 is equal to 3." 5 = 5 is read "5 is__________to 5."

is equal to

338. 5 + 2 = 7 is read "5 plus 2 ___ _______ ___ 7."

greater

339. The symbol ">" is used to represent the words "is greater than." 5 > 3 is read "5 is greater than 3." 7 > 2 is read "7 is__________ than 2."

is greater than

340. $8 > 1$ is read "8 _____ __________ _____ 1."

greater

341. $3 < 5$ means the same as $5 > 3$. That is, if 3 is less than 5, then 5 is ____________ than 3.

$>$

342. $4 < 7$ means the same as 7___4.

You could have written "is greater than," but the symbol is more convenient.

$>$

343. $8 < 15$ means the same as 15___8.

$<$

344. $25 > 20$ means the same as 20___25.

is less than

345. The symbols "<," "=," and ">" are used to express the relative order of two numbers. The symbol "<" means "_____ __________ _____."

is equal to

346. The symbol "=" means "_____ __________ _____."

is greater than

347. The symbol ">" means "_____ __________ _____."

$<$; $>$

348. The relative order of 11 and 16 could be expressed by symbols either as 11___16 or 16___11.

less than

349. If a and b are any two members of the set of natural numbers, $a < b$ means that the natural number represented by a is _____ _____ the natural number represented by b.

greater than

350. If a and b are natural numbers, $a > b$ means that the number represented by a is ____________ ____________ the number represented by b.

line graph

351. Because the natural numbers are ordered, they can be associated with points on a line segment. A line segment marked off with a common unit and used for this purpose is called a <u>line</u> <u>graph</u>. For example,

is a ________ ________ .

right

352. An arrowhead is drawn on one end of a line graph to show the direction in which the points on the line are associated with natural numbers of increasing value. Thus, the value of the numbers associated with the points on the line graph

increase from left to ___________ .

unit

353. Any unit of length may be used to represent the number 1. Thus,

are line graphs each using a different ___________ of length to represent the number 1.

scale

354. The numbers used to show the value of the unit markings on a line graph are called <u>scale</u> <u>numbers</u>. For example, on the line graph

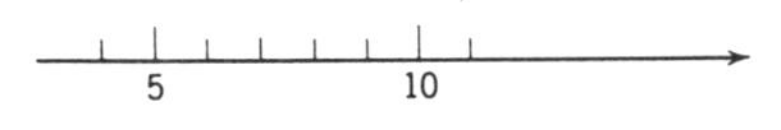

the numbers 5 and 10 are ______________ numbers.

5

355. The scale numbers on the line graph

are 25 and 50, each subdivision representing ___ units.

10

356. Each subdivision of the line graph

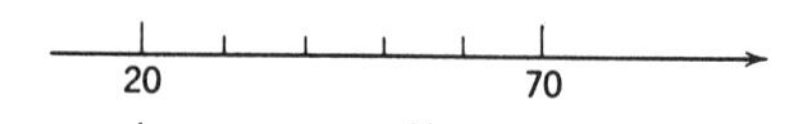

represents______ units.

45

357. Numbers may be graphed on a line graph by placing a dot on the line segment at the appropriate place. Thus,

is a graph of the numbers 30 and______.

9

358.

5 10

is a graph of the numbers 6 and____.

coordinate

359. The number corresponding to a point on a line graph is called the <u>coordinate</u> of the point. The point indicated by the dot on the line graph

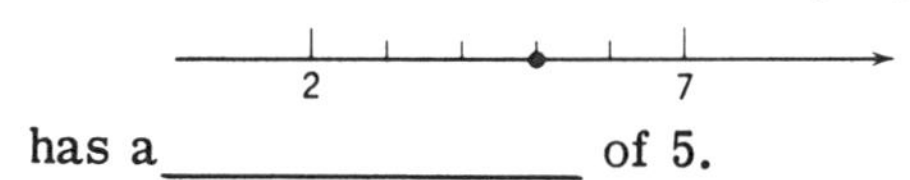

has a________________ of 5.

9

360. The coordinate of the point labeled A on the line graph

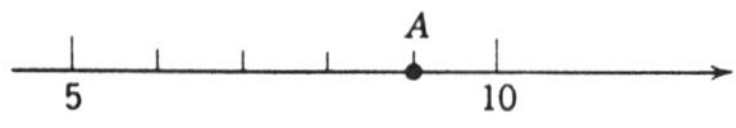

is the number_____.

coordinate

361. On the line graph

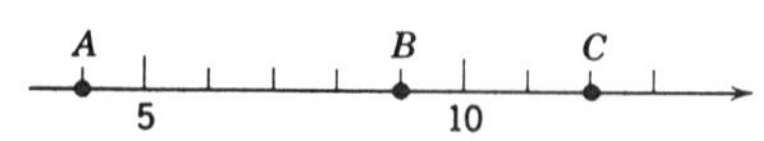

12 is the ____________________of the point labelled C.

6; 8; 11

362. The letters A, B, and C on the line graph

mark the graphs of the numbers___ , ___ , and ___.

origin

363. The point on a line graph whose coordinate is the number 0 is called the origin. Thus, the point marked A on the line graph

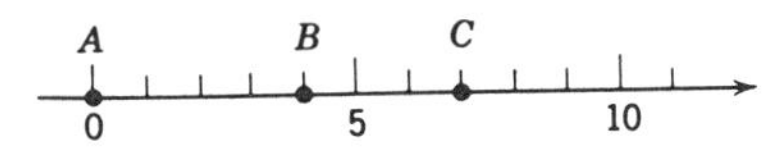

is the___________.

0

364. The coordinate of the origin on a line graph is the number_____.

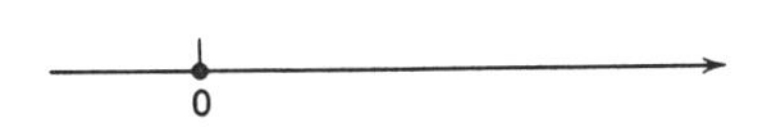

You can locate the point wherever you wish.

365. The origin can be located anywhere on the line. Graph the number 0 using the line segment below.

Perhaps you didn't use the same scale numbers that we used—there are many correct possibilities.

366. Scale the line segment and then graph the natural numbers 6, 7, and 9.

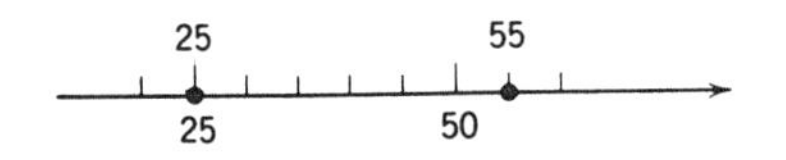

367. Mark the points corresponding to the natural numbers 25 and 55, and label these points above the line graph.

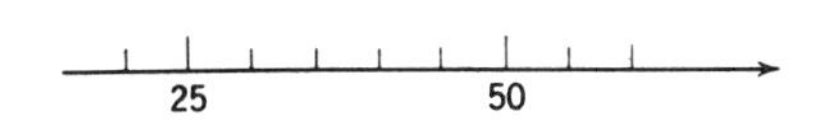

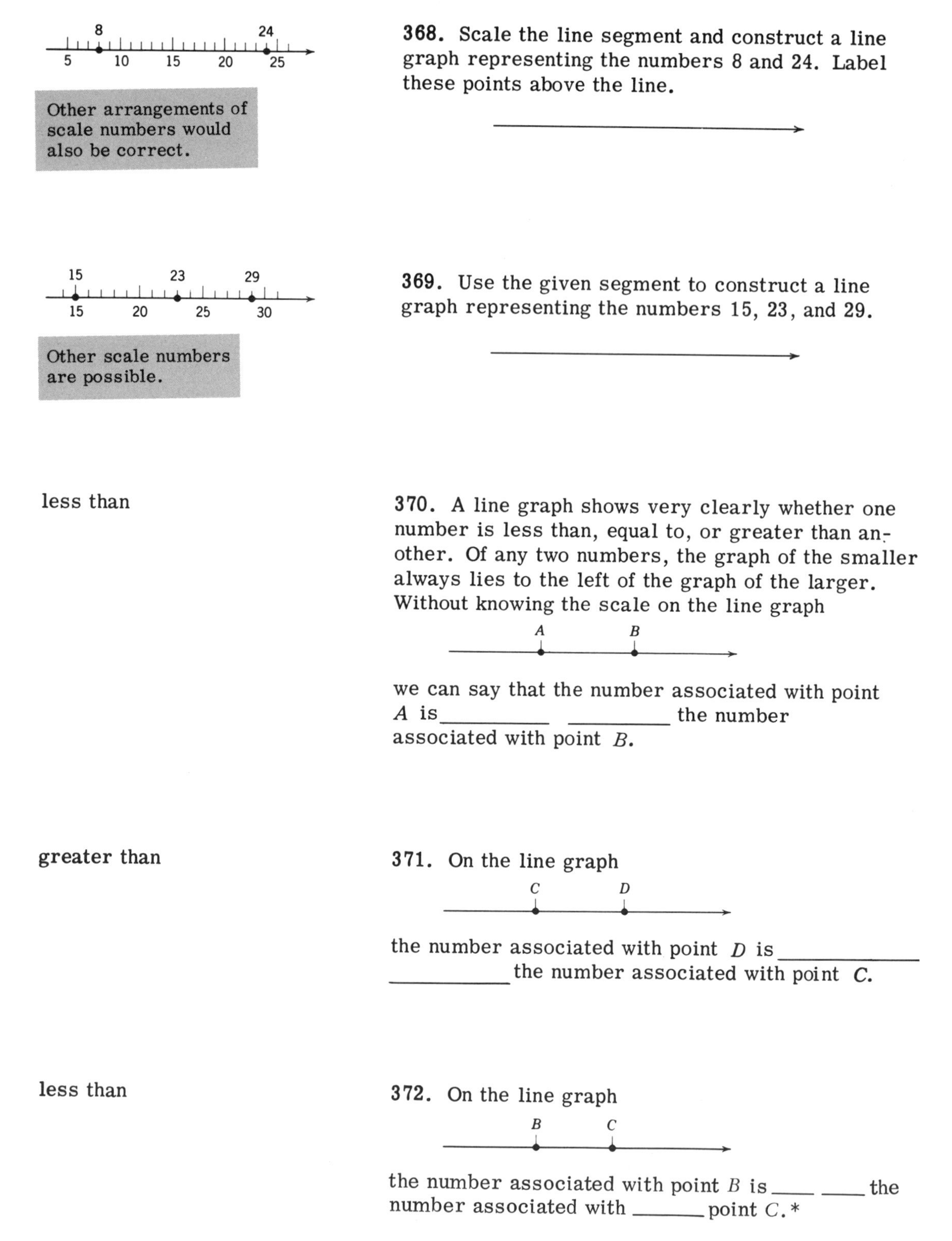

Other arrangements of scale numbers would also be correct.

368. Scale the line segment and construct a line graph representing the numbers 8 and 24. Label these points above the line.

Other scale numbers are possible.

369. Use the given segment to construct a line graph representing the numbers 15, 23, and 29.

less than

370. A line graph shows very clearly whether one number is less than, equal to, or greater than another. Of any two numbers, the graph of the smaller always lies to the left of the graph of the larger. Without knowing the scale on the line graph

we can say that the number associated with point A is __________ __________ the number associated with point B.

greater than

371. On the line graph

the number associated with point D is ______________ ___________ the number associated with point C.

less than

372. On the line graph

the number associated with point B is ____ ____ the number associated with ______ point C.*

* See Exercise 3, page 52, for additional practice.

Remark. This is all that we propose to do with line graphs at this time, but, as you will see when you continue with the next unit, these graphs offer a very concrete means of visualizing relationships between numbers.

The remainder of this unit is a review, and this last sequence of frames will give you an opportunity to obtain an overview of what you have covered. Although you should not expect to learn anything you have not learned to this point, the review will highlight the main ideas in a brief way and will reinforce what you have learned.

set

373. The word "set" refers to a collection of some kind. For example, the numbers used to count objects make up the________ of natural numbers.

natural

374. 3, 7, and 242 are members or elements of the set of__________________ numbers.

variable

375. A symbol used to represent any member of a given set is called a variable. If x is used to denote a member of the set of natural numbers, then x is a______________.

order

376. In performing the operation of addition, it is assumed that the same sum is obtained regardless of the order in which two natural numbers are added. This idea is stated:

$$a + b = b + a$$

and called the commutative law of addition. The commutative law is concerned with the___________of the terms in a sum.

grouping

377. When adding three natural numbers, a, b, and c, it is assumed that the same sum is obtained whether the addition is performed

$$(a + b) + c \text{ or } a + (b + c).$$

This assumption is called the associative law of addition. The associative law of addition is concerned with the______________of the terms in a sum.

commutative; associative

378. The addition of natural numbers is governed by two assumptions called the commutative and associative laws of addition. $3 + 5 = 5 + 3$ is an illustration of the ________________ law, and $(3 + 4) + 5 = 3 + (4 + 5)$ is an illustration of the ________________ law.

axiom

379. Assumptions in mathematics are called axioms. The commutative law of addition is an ____________.

are not

380. Since axioms are assumptions, they (are/are not) subject to proof.

difference

381. The result of subtracting one number from another is called a difference. $7 - 5$ represents the ________________ of 7 and 5.

product

382. The result of multiplying one number by another is called a product. (3)(7) represents the ________________ of 3 and 7.

factors

383. Numbers multiplied together are called factors of the product. 3 and 7 are ____________ of the product (3)(7).

product

384. The symbolic forms ab, $a \cdot b$, $(a)b$, $a(b)$, and $(a)(b)$ each represent the ____________ of the factors a and b.

order

385. It is assumed that for any two natural numbers represented by a and b, the same product is obtained regardless of the order in which the multiplication is performed. This idea is symbolized

$$ab = ba$$

and called the commutative law of multiplication. The commutative law of multiplication is concerned with the__________of the factors.

grouping

386. It is assumed that for any three natural numbers, represented by a, b, and c,

$$(ab)c = a(bc).$$

This assumption is referred to as the associative law of multiplication. The associative law of multiplication is concerned with the____________of the factors.

commutative; associative

387. Two basic laws govern the operation of multiplication. $(2)(3) = (3)(2)$ illustrates the ____________ ____________law of multiplication, and $(2 \cdot 3)4 = 2(3 \cdot 4)$ illustrates the ______________ law of multiplication.

distributive

388. The law that relates the operations of addition and multiplication is symbolized

$$a(b + c) = ab + ac$$

and is called the distributive law. The fact that $10(4 + 5) = 10(4) + 10(5)$ is an application of the ______________ law.

quotient

389. The result of dividing one number by another is called the quotient of the numbers. $\frac{7}{2}$ represents the______________when 7 is divided by 2.

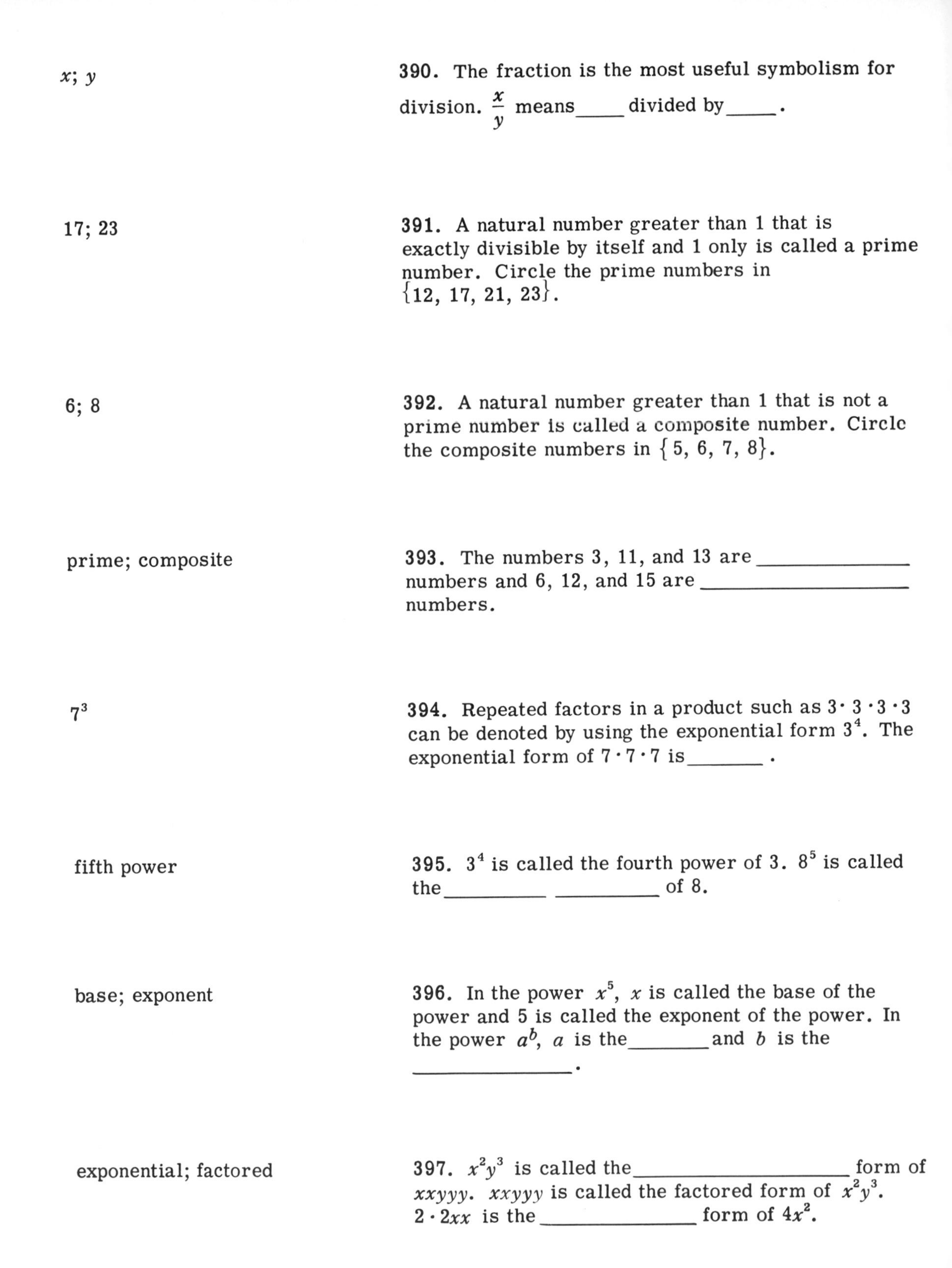

x; y

390. The fraction is the most useful symbolism for division. $\frac{x}{y}$ means _____ divided by _____.

17; 23

391. A natural number greater than 1 that is exactly divisible by itself and 1 only is called a prime number. Circle the prime numbers in $\{12, 17, 21, 23\}$.

6; 8

392. A natural number greater than 1 that is not a prime number is called a composite number. Circle the composite numbers in $\{5, 6, 7, 8\}$.

prime; composite

393. The numbers 3, 11, and 13 are ______________ numbers and 6, 12, and 15 are ______________ numbers.

7^3

394. Repeated factors in a product such as $3 \cdot 3 \cdot 3 \cdot 3$ can be denoted by using the exponential form 3^4. The exponential form of $7 \cdot 7 \cdot 7$ is _______.

fifth power

395. 3^4 is called the fourth power of 3. 8^5 is called the __________ __________ of 8.

base; exponent

396. In the power x^5, x is called the base of the power and 5 is called the exponent of the power. In the power a^b, a is the ________ and b is the ______________.

exponential; factored

397. x^2y^3 is called the ____________________ form of $xxyyy$. $xxyyy$ is called the factored form of x^2y^3. $2 \cdot 2xx$ is the _______________ form of $4x^2$.

polynomial

"trinomial" is also correct.

398. An expression made up of terms such as $2x$, $4x^3$, and 5, where each term involves the multiplication of numerals and variables only, is called a polynomial. $y^3 + 2y + 3$ is a ________________.

binomial

399. Any polynomial containing just one term is called a monomial, a polynomial containing two terms is called a binomial, and a polynomial containing three terms is called a trinomial. $3x - y$ is a ______________.

monomial; binomial; trinomial

400. $4x^2$ is a ____________, $2x + 4$ is a ____________, and $x^2 + 2x + 1$ is a ____________.

3

401. The numerical factor of a term is called the numerical coefficient of the term. The numerical coefficient of $3x^2y$ is____.

$3x^2$; x^2

402. Terms that are identical in their variable factors are called like terms. Circle the like terms in the polynomial $3x^2 + 2x + x^2 + 5$.

$4x + 3y$

403. The distributive law in the form

$$ab + ac = a(b + c)$$

justifies writing the sum of two products $3x + 4x$ as $(3 + 4)x$ or $7x$. The application of the distributive law in this form is commonly referred to as combining or adding like terms. The result of combining the like terms in the polynomial $3x + 2y + x + y$ is

____________.

$8x + 5y$

404. Combining or adding like terms in an expression is also referred to as simplifying the expression. The result of simplifying $2x + 5y + 6x$ is__________.

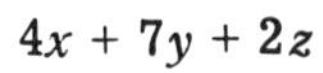

405. **The result of simplifying $(x + 2y + z) +$ $(3x + 5y + z)$ is ______________.**

less than

406. The symbols "<," "=," and ">" mean "is less than," "is equal to," and "is greater than," respectively. $x < 5$ means that x is ______ ______ 5.

right

407. Line graphs are useful in visualizing the relative order of two numbers. Of any two numbers, the graph of the smaller will lie to the left of the graph of the larger. On a line graph, the graph of 11 will lie to the (left/right) of the graph of 7.

Remark. This concludes Unit I. To see what you have learned, you can take one of the self-evaluation tests on pages 53-54. If you completed one form of the test before you started the program, use the alternate form now.

EXERCISES AND ANSWERS

Exercise 1. If you have difficulty with this exercise, reenter the program at Frame 170.

Write in completely factored form.

1. 10 **2.** 15 **3.** 18 **4.** 30 **5.** $40m^2$ **6.** $27n^3$ **7.** $32x^2y$ **8.** $12a^3b$

9. $48p^3q^2$ **10.** $36n^4s^2$

Write in exponential form.

11. $3 \cdot 3 \cdot 3$ **12.** $5 \cdot 5 \cdot 5 \cdot 5$ **13.** $2 \cdot x \cdot x \cdot x \cdot x$ **14.** $7 \cdot y \cdot y \cdot y \cdot y \cdot y$

15. $11 \cdot x \cdot x \cdot y$ **16.** $13 \cdot m \cdot m \cdot m \cdot n$ **17.** $x \cdot x \cdot x \cdot y \cdot y \cdot y \cdot y$

18. $17 \cdot t \cdot t \cdot t \cdot t \cdot t \cdot s$ **19.** $7 \cdot a \cdot a \cdot b \cdot b \cdot b \cdot c$ **20.** $2 \cdot x \cdot x \cdot x \cdot y \cdot y \cdot y \cdot z \cdot z$

Answers

1. $2 \cdot 5$ **2.** $3 \cdot 5$ **3.** $2 \cdot 3 \cdot 3$ **4.** $2 \cdot 3 \cdot 5$ **5.** $2 \cdot 2 \cdot 2 \cdot 5 \cdot m \cdot m$

6. $3 \cdot 3 \cdot 3 \cdot n \cdot n \cdot n$ **7.** $2 \cdot 2 \cdot 2 \cdot 2 \cdot 2 \cdot x \cdot x \cdot y$ **8.** $2 \cdot 2 \cdot 3 \cdot a \cdot a \cdot a \cdot b$

9. $2 \cdot 2 \cdot 2 \cdot 2 \cdot 3 \cdot p \cdot p \cdot p \cdot q \cdot q$ **10.** $2 \cdot 2 \cdot 3 \cdot 3 \cdot n \cdot n \cdot n \cdot n \cdot s \cdot s$ **11.** 3^3

12. 5^4 **13.** $2x^4$ **14.** $7y^5$ **15.** $11x^2y$ **16.** $13m^3n$ **17.** x^3y^4 **18.** $17t^5s$

19. $7a^2b^3c$ **20.** $2x^3y^3z^2$

Exercise 2. If you have difficulty with this exercise, reenter the program at Frame 272.

Simplify.

1. $x + 2x + 3x$ **2.** $2y + y + y$ **3.** $2x + z + 2x$ **4.** $5a + 2b + 3a$ **5.** $3c + 2d + 2c + 3d$

6. $4x + y + 2x + 3y$ **7.** $x^2 + 2x + 2x^2 + 3x$ **8.** $4z^2 + 2z^2 + 3z + z^2$ **9.** $(a^2 + 3a) + (2a + a^2)$

10. $(b^2 + 3b) + (2b^2 + 5b)$ **11.** $(z^2 + 2z + 3) + (3z^2 + 2z + 1)$

12. $(n^2 + n + 1) + (2n^2 + 3n + 5)$ **13.** $(m^2 + 3) + (4m^2 + 2m)$ **14.** $(2p^2 + 3p) + (4p^2 + 2p + 1)$

15. $(3m^2n + 2m^2n) + (5m^2n + m^2n)$ **16.** $(6abc + 4abc) + (abc + 3abc)$

Answers

1. $6x$ **2.** $4y$ **3.** $4x + z$ **4.** $8a + 2b$ **5.** $5c + 5d$ **6.** $6x + 4y$ **7.** $3x^2 + 5x$

8. $7z^2 + 3z$ **9.** $2a^2 + 5a$ **10.** $3b^2 + 8b$ **11.** $4z^2 + 4z + 4$ **12.** $3n^2 + 4n + 6$

13. $5m^2 + 2m + 3$ **14.** $6p^2 + 5p + 1$ **15.** $11m^2n$ **16.** $14abc$

Exercise 3. If you have difficulty with this exercise, reenter the program at Frame 332.

Replace the , with one of the symbols $<$, $=$, or $>$ to show the order of the given numbers.

1. 3 , 7 **2.** 6 , 4 **3.** 3 + 2 , 5 **4.** 7 + 1 , 7 **5.** 4 , 3 + 2 **6.** 12 , 6 + 6

7. $2 \cdot 3$, 4 + 1 **8.** $4 \cdot 3$, 6 + 2 **9.** $6 \cdot 1$, $3 \cdot 2$ **10.** $5 \cdot 4$, $7 \cdot 3$

Construct a line graph representing the given numbers. Use any appropriate scale.

11. 1, 3, 5 **12.** 2, 4, 6 **13.** 1, 4, 5 **14.** 2, 3, 7 **15.** 11, 12, 16 **16.** 6, 8, 14

17. 25, 29, 36 **18.** 43, 47, 50 **19.** 104, 105, 106 **20.** 93, 97, 99

Answers

1. $3 < 7$ **2.** $6 > 4$ **3.** $3 + 2 = 5$ **4.** $7 + 1 > 7$ **5.** $4 < 3 + 2$ **6.** $12 = 6 + 6$

7. $2 \cdot 3 > 4 + 1$ **8.** $4 \cdot 3 > 6 + 2$ **9.** $6 \cdot 1 = 3 \cdot 2$ **10.** $5 \cdot 4 < 7 \cdot 3$

11. 0 5 **16.** 5 10 15

12. **17.**

13. **18.**

14. **19.**

15. **20.**

SELF-EVALUATION TEST, FORM A

1. Circle the two natural numbers that are members of $\{3, 10.5, 6/5, 171\}$.
2. A symbolic representation of the________________ law of addition is given by $a + b = b + a$.
3. A symbolic representation of the________________ law of multiplication is given by $(ab)c = a(bc)$.
4. Write the product of 5 and $a + b$ as a sum.
5. The expression $a - b$ represents the________________ of a and b.
6. A symbolic representation of the________________ law is given by $a(b + c) = ab + ac$.
7. Write the quotient of $a + b$ divided by 3 in fractional form.
8. Circle the two prime numbers in the set $\{3, 8, 12, 23\}$.
9. The expression $21x^3y^2$ appears in completely factored form as ________________.
10. x^4 is called a__________ of x.
11. In the expression 4^5, the symbol "5" is called an ____________.
12. Express $aabbb$ in exponential form.
13. A polynomial containing two terms is called a______________.
14. What is the numerical coefficient of $7x^3y^2$?
15. Simplify: $2x^2y + xy^2 + 3xy^2$.
16. Simplify: $3x^2y + y^2x + x^2y$.
17. Add: $(3y^2 + y + 2) + (y^2 + 4y + 1)$.
18. $x < y$ means that the number represented by y is (less than/greater than) the number represented by x.
19. Using the symbol "$>$," the relative order of 17 and 21 can be expressed__________.
20. On the line graph

R S

the number associated with point S is (less than/greater than) the number associated with point R.

SELF-EVALUATION TEST, FORM B

1. Circle the two natural numbers that are members of $\{4.1, 17, 23/2, 109\}$.
2. A symbolic representation of the ________________ law of addition is given by $a + (b + c) = (a + b) + c$.
3. A symbolic representation of the ________________ law of multiplication is given by $ab = ba$.
4. Write the product of 3 and $x + y$ as a sum.
5. The expression $x - y$ represents the ________________ of x and y.
6. A symbolic representation of the ________________ law is given by $a(b + c) = ab + ac$.
7. Write the quotient of $x + 2y$ divided by 5 in fractional form.
8. Circle the two prime numbers in the set $\{5, 10, 17, 21\}$.
9. The expression $35x^2y^3$ appears in completely factored form as ________________.
10. y^3 is called a __________ of y.
11. In the expression 7^5, the symbol "7" is called the ________.
12. Express $bbccc$ in exponential form.
13. A polynomial containing three terms is called a ________________.
14. What is the numerical coefficient of $5x^2y^4$?
15. Simplify: $3a^2bc + 4ab^2c + 6ab^2c$.
16. Simplify: $2a^2b + 3ab^2 + a^2b$.
17. Add: $(3y^2 + 2y + 1) + (2y^2 + y + 3)$.
18. $a < b$ means that the number represented by a is (less than/greater than) the number represented by b.
19. Using the symbol "<," the relative order of 23 and 15 can be expressed __________.
20. On the line graph

M N

the number associated with point M is (less than/greater than) the number associated with point N.

ANSWERS TO TESTS

Form A

1. 3; 171
2. commutative
3. associative
4. $5a + 5b$
5. difference
6. distributive
7. $\frac{a + b}{3}$
8. 3; 23
9. $3 \cdot 7xxxyy$
10. power or fourth power
11. exponent
12. a^2b^3
13. binomial
14. 7
15. $2x^2y + 4xy^2$
16. $4x^2y + y^2x$
17. $4y^2 + 5y + 3$
18. greater than
19. $21 > 17$
20. greater than

Form B

1. 17; 109
2. associative
3. commutative
4. $3x + 3y$
5. difference
6. distributive
7. $\frac{x + 2y}{5}$
8. 5; 17
9. $5 \cdot 7xxyyy$
10. power or cube or third power
11. base
12. b^2c^3
13. trinomial
14. 5
15. $3a^2bc + 10ab^2c$
16. $3a^2b + 3ab^2$
17. $5y^2 + 3y + 4$
18. less than
19. $15 < 23$
20. less than

VOCABULARY, UNIT I

The frame in which each word is introduced is shown in parentheses.

associative law of addition (61)
associative law of multiplication (120)
axiom (40)
base (215)
binomial (252)
braces (3)
brackets (59)
coefficient (261)
combining like terms (308)
commutative (38)
commutative law of addition (44)
commutative law of multiplication (109)
completely factored form (196)
composite number (180)
coordinate (359)
difference (90)
distributive law (144)
element (5)
equals (26)
exponent (213)
exponential form (228)
expression (236)
factored form (187)
factors (97)
finite set (18)
fraction (154)
infinite set (20)
grouping (67)
like terms (272)
line graph (351)
member (4)
monomial (250)
natural numbers (8)
numerical coefficient (264)
order (66)
ordered set (332)
origin (363)
parentheses (57)
polynomial (243)
power (217)
prime factors (190)
prime number (173)
product (96)
quotient (150)
replacement set (29)
scale numbers (354)
set (1)
simplifying expressions (316)
substitution axiom (82)
sum (34)
symmetric law of equality (77)
term (238)
trinomial (257)
variable (24)

UNIT II Integers

OBJECTIVES

Upon completion of the unit the student should:

1. Know what is meant by "integer," and be able to identify representative members of the set of integers.

2. Be able to graph integers.

3. Know the meaning of "absolute value," and be able to identify $|x|$ for representative integers x.

4. Be able to perform routine operations using integers.

5. Be able to simplify expressions involving numerals and variables representing integers, including expressions involving several operations.

6. Be able to numerically evaluate expressions for integral values of the variables in the expression.

CONTENTS

UNIT II
INTEGERS

Remark. In Unit I, we worked only with the set of natural numbers, $\{1, 2, 3, \ldots\}$, and with variables whose replacement set was the set of natural numbers. In this unit, we will be working with an enlargement of this set. Since the line graphs we used to visualize natural numbers will also serve to visualize the new set of numbers, the concern of the first portion of this unit will be to review briefly the concept of a line graph and then to use the line graph to introduce some new numbers.

line graph

1. A line graph can be used to visualize the relative order of numbers. For example,

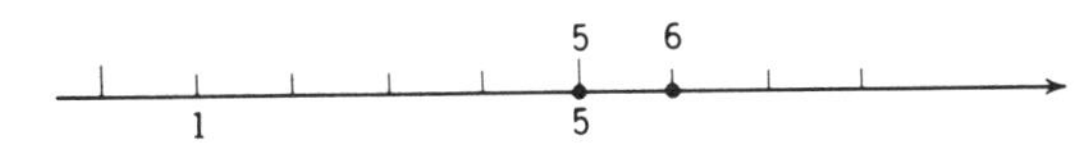

is a_______ _______ showing the relative order of the numbers 5 and 6.

one

2. A line graph is constructed by drawing a line segment with an arrowhead on the right end and then scaling the segment as desired. Thus,

is a line graph where each division represents______ units.

less

"smaller" would do.

3. A line graph shows very clearly whether one number is less than, equal to, or greater than another number. Of any two numbers, the graph of the smaller always lies to the left of the graph of the larger. Thus, on the line graph

A B

the number corresponding to the point A is ________ than the number corresponding to the point B.

greater

"larger" would do.

4. On the line graph

C D

the number associated with the point D is ________ than the number associated with the point C.

origin

5. Recall that the point on a line graph associated with the number 0 is called the origin. Thus, on the line graph

A B
1 5

the point labeled A is the ________.

coordinate

6. The coordinate of a point on a line graph is the number associated with the point. Thus, 0 is the ________ of the origin.

0

Or zero.

7. The coordinate of the origin on a line graph is the number____.

3

8. On a line graph, each unit distance to the right of the origin is associated with a natural number. On the line graph

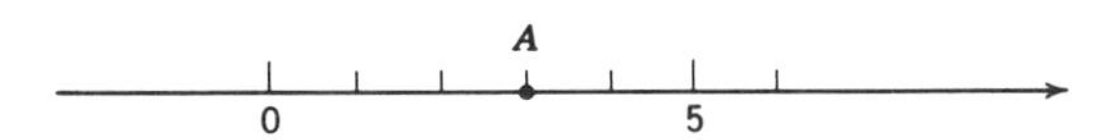

the point labeled A is associated with the natural number____.

29

9. On the line graph

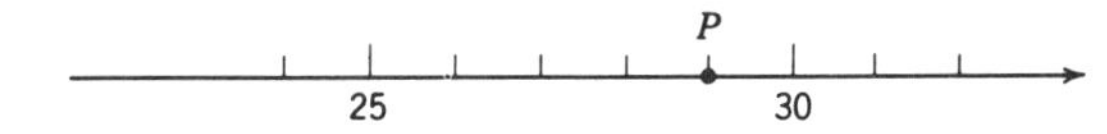

the point labeled P is associated with the natural number____.

right

10. Any natural number corresponds to a point located to the right of the origin. Thus, on a line graph, the graph of the number 4 lies to the________ of the origin.

left

11. If a line graph is drawn extending in both directions from 0, for every point to the right of 0 there is a point located the same distance to the left of 0. On the line graph

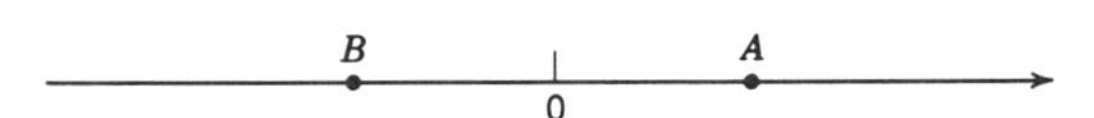

the point labeled A is to the right of the origin and the point labeled B is to the________of the origin.

negative

12. For each natural number corresponding to a point to the right of the origin there is another number corresponding to a point located the same distance to the left of the origin called the negative of the first number. Thus, if the points C and D are the same distance from the origin on the line graph

D 0 C

the number associated with the point D is the __________ of the number associated with the point C.

negative

13. A number that is the negative of a natural number is represented by a numeral preceded by a minus sign. Thus, -2 represents the negative of 2; -3 represents the negative of 3; and -5 represents the __________ of 5.

negative

14. The number -5 is read "negative five." -8 is read "__________ eight."

negative

15. -27 is read "__________ twenty-seven."

negative

16. Numbers such as -3, -5, and -11 are called negative numbers. -7 is a __________ number.

negative

17. -8 and -107 are __________ numbers.

-4

18. For every natural number, a, there corresponds a negative number, $-a$. Corresponding to the natural number 4 is the negative number____.

four

19. The graphs of 4 and -4 appear as

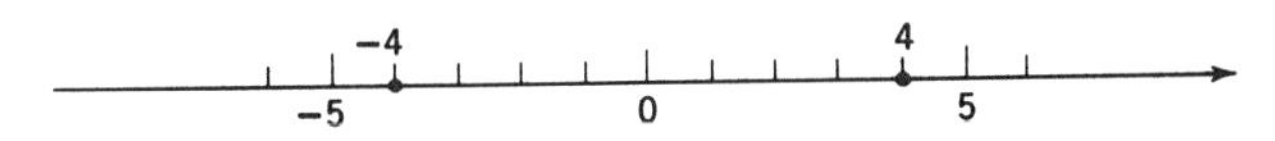

and each is located________ units from the origin, but in opposite directions.

-6

20. The graphs of 6 and the negative of 6 appear as

The negative of 6 is represented by the numeral_____.

15

21. If two numbers correspond to points on a line graph that are located the same distance from the origin but in opposite directions, each is called the negative of the other. Thus, the negative of -11 is 11. Similarly, the negative of -15 is_____.

3

22. Since -5 and 5 are represented by points the same distance from the origin, each is the negative of the other. Thus, the negative of -5 is 5. Similarly, the negative of -3 is____.

7

23. The negative of -7 is_____.

-8

24. 8 is the negative of_____.

negative

25. 27 is the____________ of -27.

18

26. In symbols, the negative of -3 can be written $-(-3)$, and read "the negative of negative 3." $-(-3) = 3$. Similarly, $-(-18) =$_____.

29

27. $-(-29) =$ _____ .

positive

28. To distinguish between negative numbers and numbers that are not negative numbers, the term positive number is used. Any number corresponding to a point on a line graph that lies to the right of the origin is a ____________ number.

positive

29. The sign + is sometimes prefixed to the symbol representing a positive number. Thus, +8, +10, and +17 represent ____________ numbers.

positive

30. If no sign is prefixed to a symbol representing a number, the symbol is understood to represent a positive number. Thus, both the symbols +3 and 3 represent the same ____________ number.

4

31. The symbols +4 and _____ represent the same positive number.

+7

32. The symbols _____ and 7 represent the same positive number.

negative

33. If x represents a positive number, $-x$ represents a ____________ number.

negative

34. If y represents a positive number, $-y$ represents a ____________ number.

positive

The negative of a negative number is a positive number.

35. If x represents a negative number, $-x$ (the negative of x) represents a ____________ number.

positive

36. If b represents a negative number, $-b$ represents a__________number.

$-c$

37. If c represents a negative number, the negative of c is written_____and represents a positive number.

positive

38. The symbol $-x$ may represent either a positive or a negative number. If x represents a positive number, $-x$ represents a negative number, but if x represents a negative number, $-x$ represents a__________ number.

positive

39. -3 always represents a negative number. $-x$, however, may represent either a negative or a __________number, depending upon whether x is positive or negative.

signed

40. The set of all natural numbers and their negatives is an example of a set of signed numbers. The numbers -8, 5, 7, and -6 are__________numbers.

origin

41. Each signed number corresponds to a point located in one or the other direction from the__________on a line graph.

signed

42. -8, 3, -2, and 5 are__________numbers.

negative

43. The number 0 corresponds to the point on a line graph at the origin and hence has no sign associated with it. 0 is neither a positive number nor a __________ number.

positive

44. The number 0 separates the set of positive numbers from the set of negative numbers. Any number greater than 0 (i.e., corresponding to a point to the right of 0 on a line graph) is a ____________ number.

negative

45. Any number that is less than 0 (i.e., corresponding to a point to the left of 0 on a line graph) is a ____________ number.

integer

46. The set of all natural numbers and their negatives, together with 0, is called the set of <u>integers</u>. Thus, 3, −3, and 0 are integers. −7 is an ____________.

integer

47. −25 is an ____________.

integer

"signed number" is correct but it doesn't fit here.

48. An integer can be described as being any positive or negative whole number or 0. Because −8 is a negative whole number, −8 is an ____________.

integer

49. 143 is a natural number. Therefore, it is an ____________.

is

50. 0 (is/is not) an integer.

integer

51. All natural numbers are integers. If x is a natural number, then x must be an ____________.

natural

52. Not all integers are natural numbers. For example, −7 is an integer, but −7 is not a ____________ number.

natural

53. "Natural numbers" and "positive integers" mean the same thing. Any positive integer is a ____________ number and vice versa.

left

54. Recall that the natural numbers are ordered. If x and y represent natural numbers, x is either greater than y, equal to y, or less than y. If x is less than y, then the point corresponding to x on a line graph will lie to the_______ of the point corresponding to y.

less

55. Because the point corresponding to a is to the left of the point corresponding to b on the line graph

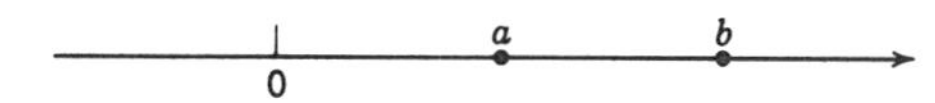

the natural number represented by a is_______than the natural number represented by b.

left

56. The integers are ordered. That is, if x and y represent integers, x is either greater than y, equal to y, or less than y. If x is less than y, then the point corresponding to x on a line graph will lie to the _____ of the point corresponding to y.

less

57. Because the point corresponding to a is to the left of the point corresponding to b on the line graph

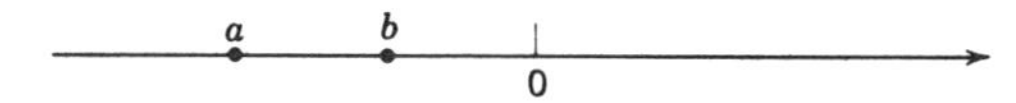

the integer represented by a is______than the integer represented by b.

greater

The graph of q is to the right of the graph of t.

58. On the line graph

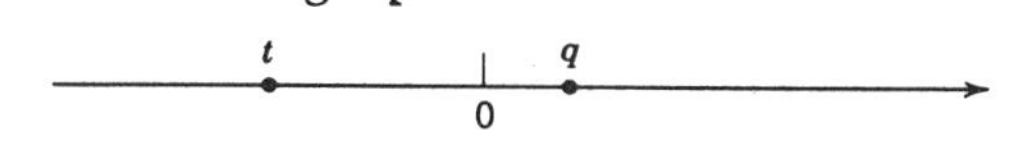

the integer represented by q is __________ than the integer represented by t.

less; left

59. The graphs of the numbers -8 and -6 are shown on the line graph

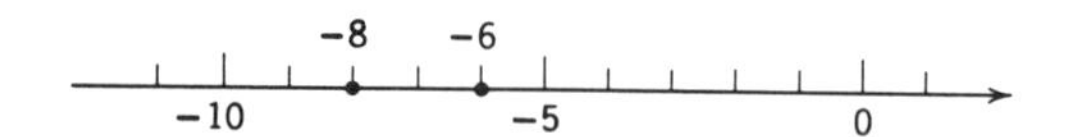

It is clear from the line graph that -8 is ________ than -6 because the point corresponding to -8 lies to the______ of the point corresponding to -6.

less

60. The line graph

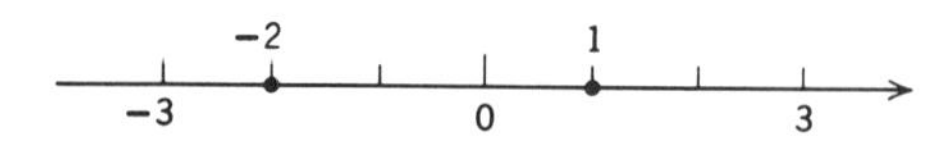

shows that -2 is______ than 1.

61. Graph the numbers -8, -3, and 2 on the line graph

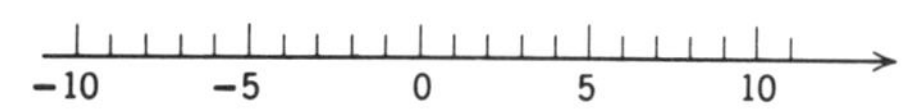

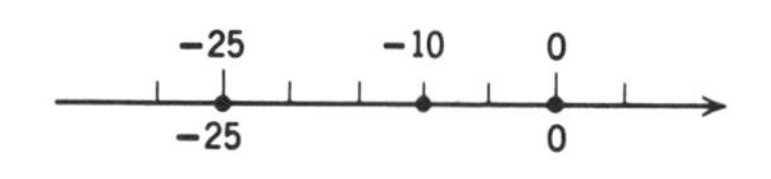

62. Graph the numbers -25, -10, and 0 on the line graph

-16; -5

63. Recall that the symbol "<" means "is less than". The relative order of the numbers -6 and -4 can be shown by writing $-6 < -4$. The relative order of -16 and -5 can be shown by writing____ < _____.

-6; 1

64. The relative order of the integers -6 and 1 can be shown by writing____ < _____.

greater

65. If -6 is less than 1, then 1 is____________ than -6.

$1 > -6$

66. $-6 < 1$ means the same as ____ > ____.

$-18; -1; 0$

67. The relative order of the numbers -1, -18, and 0 can be shown by writing ____ < ____ < ____.

absolute

68.* The numbers 4 and -4 correspond to points located the same distance from the origin, but in opposite directions. To refer only to the distance of these points from the origin, and not to their direction, the symbols $|-4|$ and $|4|$ are used. The symbol $|-4|$ is read "the absolute value of negative four." $|-2|$ is read "the ____________ value of negative 2."

absolute

69. $|4|$ is read "the absolute value of four." $|7|$ is read "the ____________ value of seven."

8

70. The absolute value of 0 is 0. The absolute value of any number other than 0 is positive. Thus, $|0| = 0$, $|4| = 4$, and $|-4| = 4$. $|-8| =$ ____.

5

71. $|-5| =$ ____.

6

72. $|6| =$ ____.

86

73. $|-86| =$ ____.

18

74. The absolute value of -18 is ____.

17

75. The absolute value of 17 is ____.

is

76. $|7|$ (is/is not) equal to $|-7|$.

Both represent 7.

is

77. $|-6|$ (is/is not) equal to 6.

is not

78. $|-3|$ (is/is not) equal to -3.

The absolute value of any number other than 0 is positive.

0

79. $|0| =$ ____.

Remark. We have now extended the set of natural numbers to the set of integers. We can graph integers on line graphs and we have defined what we mean by the absolute value of an integer in terms of a distance from the origin on a line graph.

We shall next turn our attention to performing operations with integers. Since the operation of addition is perhaps the most basic one, we shall study it first.

counting; right

80. Positive integers can be associated with the counting of units to the right on a line graph. Thus, the number 5 is associated with __________ five units to the __________ from the origin on a line graph.

right; origin

81. The line graph

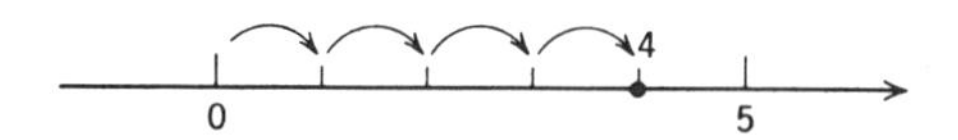

shows the counting of four units to the right of the origin, and the point arrived at is labeled 4. The intege 4 can be associated with counting four units to the ________ from the __________.

6

Or 4 + 2 if you like.

82. The sum 4 + 2 can be thought of as the number associated with the point arrived at after counting off four units to the right from the origin, and then, from this point, counting off two more units to the right.

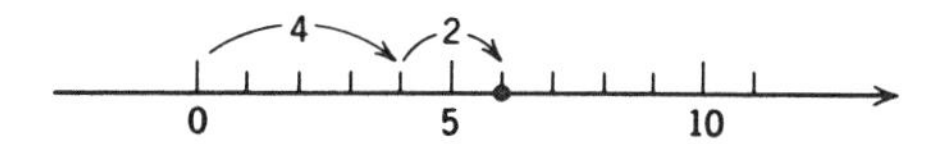

The number associated with the point arrived at last is____.

11

Or 6 + 5 if you like.

83. The line graph

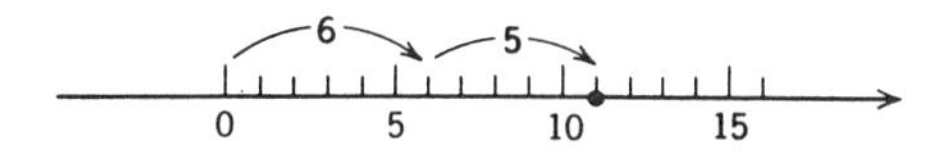

shows the graphical addition of the two positive integers 6 and 5. First, six units are counted off to the right from the origin, then, from this point, five more units are counted off to the right. The number associated with the point arrived at last is_____.

84. Using the line graph in the preceding frame for a model, show the addition of 5 and 3 on the line graph

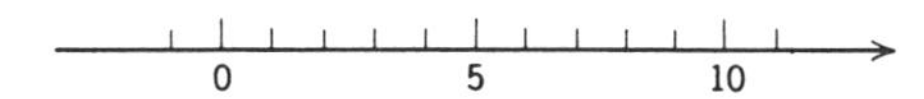

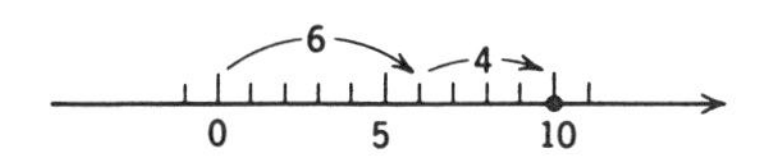

85. Show the addition of 6 and 4 on the line graph

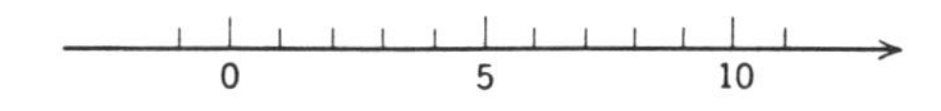

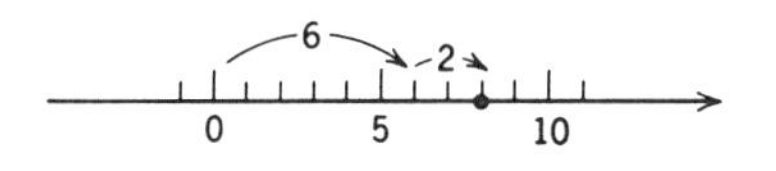

86. Show the addition of 6 and 2 on the line graph

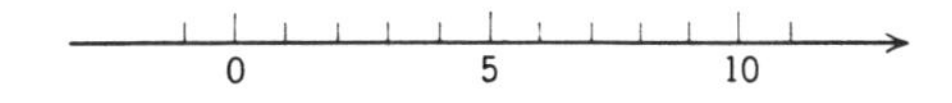

counting; left

87. Negative numbers can be associated with counting to the left on a line graph. Thus, the number -3 is associated with __________ three units to the _________ from the origin on a line graph.

left; origin

88. The line graph

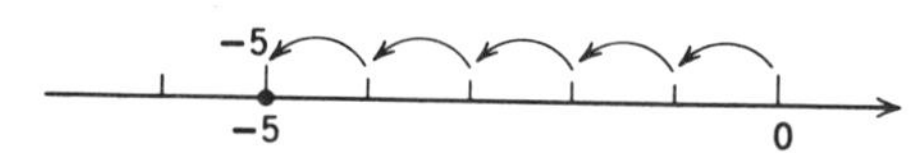

shows the counting of five units to the left from the origin, and the point arrived at is labeled -5. Thus, the integer -5 can be associated with counting five units to the_______ from the___________.

-8

Or $(-5) + (-3)$.

89. To find the sum of -5 and -3, five units can first be counted off to the left from the origin, and, from this point, three more units can be counted off to the left. The number associated with the point arrived at last is_____.

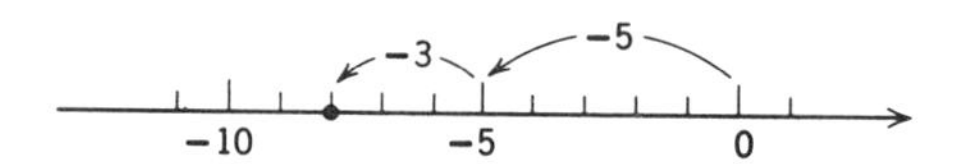

-11

Or $(-4) + (-7)$.

90. The line graph

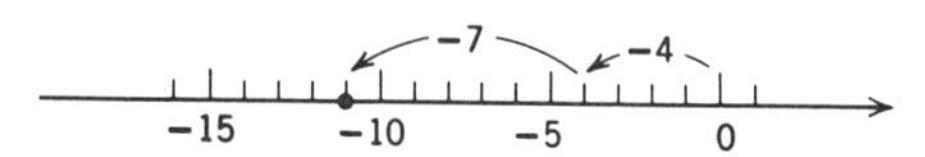

shows the graphical addition of the two negative integers -4 and -7. First, four units are counted off to the left of the origin, then, from this point, seven more units are counted off to the left. The number associated with the point arrived at last is______.

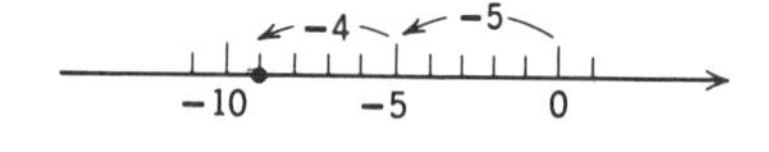

91. Show the addition of -5 and -4 on the line graph

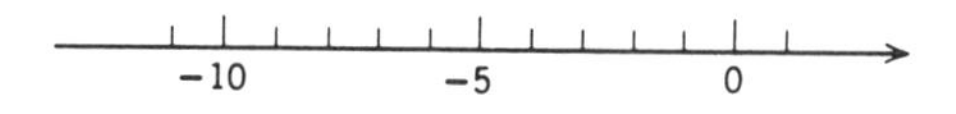

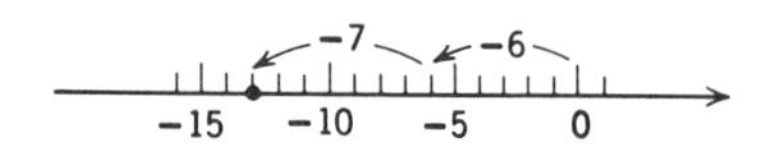

92. Show the addition of -6 and -7 on the line graph

negative

93. It can be seen from the preceding examples that the sum of two positive integers is a positive integer and that the sum of two negative integers is a ____________integer.

-18

94. It is not necessary to draw a line graph to perform the addition of two integers. The sums may be calculated mentally by imagining the result of moving to the left or right on a line graph. The sum of -8 and -10 is______.

-17

95. Add -3 and -14.

sum

96. The expression $(-2) + (-7)$ is read "negative two plus negative seven," and represents the______ of -2 and -7.

-9

97. $(-2) + (-7) =$_____.

-28

98. $(-3) + (-25) =$_____.

23

99. $(+10) + (+13) =$_____.

-33

100. $(-18) + (-15) =$_____.

9

101. The integer 0 plays a special role in addition. If x is an integer, then $x + 0 = x$. That is, the addition of 0 to a given number produces the given number as a sum. $3 + 0 = 3$, $-8 + 0 = -8$, $9 + 0 =$_____.

18

102. $18 + 0 =$____.

−18

103. $0 + (-18) =$ ____.

y

104. $0 + y =$ ____.

−4

105. It is assumed that the commutative law of addition is true for integers. Thus,

if a *and* b *are integers,*

$$a + b = b + a.$$

This means that $(-4) + (-7)$ is the same as $(-7) + ($___$)$. Each sum is equal to −11.

commutative

Remember that the commutative law is concerned with the order of the terms.

106. $(-3) + (-9) = (-9) + (-3)$ is an example of an application of the ________ law of addition for integers.

−11

107. The commutative law of addition for integers assures us that $(-11) + (-20) = (-20) + ($____$)$.

$[(-3) + (-4)]$

108. It is assumed that the associative law of addition is true for integers. Thus,

if a, b, *and* c *are integers,*

$$(a + b) + c = a + (b + c).$$

This means that $[(-2) + (-3)] + (-4)$ is the same as $(-2) + [\quad + \quad]$. Each sum equals −9.

associative

Remember that the associative law is concerned with the grouping of the terms.

109. $[(-5) + (-2)] + (-3) = (-5) + [(-2) + (-3)]$ is an example of the ________ law of addition.

[(− 11) + (− 5)]

110. The associative law of addition assures us that (− 15) + (− 11) + (− 5) equals either [(− 15) + (− 11)] + (− 5) or (− 15) + [___ + ___]. Each sum is equal to −31.

12

111. 5 + 4 + 3 equals either (5 + 4) + 3 or 5 + (4 + 3). Each sum is equal to_____.

− 13

112. (−2) + (− 4) + (− 7) = _____.

13

113. 7 + 0 + 6 = _____.

− 18

114. (− 10) + (− 8) + (0) = _____.

Remark. We have thus far restricted ourselves to the addition of two integers of like sign. Our next task will be to learn to add integers of opposite sign. We have seen that when we add a positive number to a positive number we obtain a positive number, and when we add a negative number to a negative number,we obtain a negative number. When we add numbers of opposite sign, however, sometimes we obtain a positive number and sometimes a negative number, depending upon the relative absolute value of the numbers being added.

2

Or (7) + (−5).

115. The sum of any positive and any negative integer can be thought of in terms of counting on a line graph. For example, the sum (7) + (− 5) can be thought of as the number associated with the point arrived at last after counting seven units to the right from the origin, and then, from this point, counting five units back to the left.

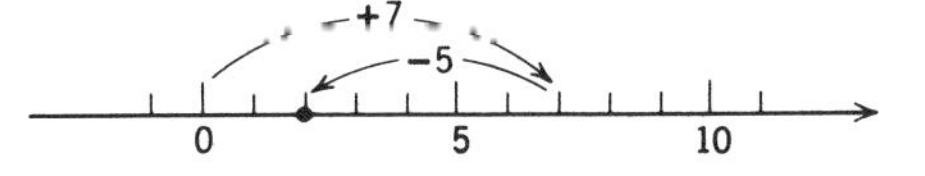

The number associated with the point arrived at by this process is___.

5; 5

116. The sum $(8) + (-3)$ can be thought of as the number associated with the point arrived at last by first counting eight units to the right from the origin on a line graph, and then, from this point, counting three units back to the left.

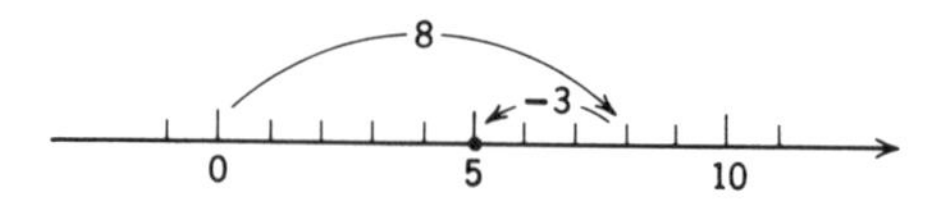

The number associated with the point arrived at last is____. Thus, the sum of 8 and -3 is____.

5

(−3) + (8) must be the same as (8) + (−3).

117. Since $(8) + (-3) = 5$, then, by the commutative law of addition, $(-3) + (8) =$____.

-5; -5

118. To find the sum of -8 and 3, eight units can be counted off to the left from the origin on a line graph and then, from this point, three units can be counted off to the right.

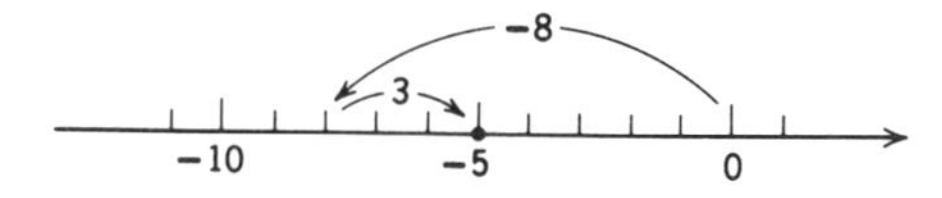

The number associated with the point arrived at last is____. Thus, the sum of -8 and 3 is____.

-5

119. Since $(-8) + (3) = -5$, $(3) + (-8) =$____.

-2

Eight units to the left and six units to the right.

120. Of course, to find a sum such as $(-8) + (6)$, it is not necessary to go through a counting process on an actual line graph. Such sums can be computed mentally by imagining the result of moving to the right and left (or left and right) along the number line. Thus, $(-8) + (6) =$____.

-1 **121.** $(-3) + (2) =$ ____.

1 **122.** $(5) + (-4) =$ ____.

-2 **123.** $(7) + (-9) =$ ____.

3 **124.** $(8) + (-5) =$ ____.

Remark. Notice how we are using parentheses here, to help identify whether a number is positive or negative. There is no difference between writing $3 + 5$ and $(3) + (5)$, but now that we are working with both positive and negative numbers, it is very convenient to be able to look at $(3) + (5)$ and see immediately that positive 3 is to be added to positive 5. Similarly, when we write $(-3) + (5)$ instead of $-3 + 5$, it is simply to make it clear that we are adding negative 3 to positive 5.

14 **125.** $(-5) + (19) =$ ____.

-16 **126.** $(3) + (-19) =$ ____.

0 **127.** $(-5) + (+5) =$ ____.

0 **128.** $(+7) + (-7) =$ ____.

0 **129.** $(x) + (-x) =$ ____, where x represents an integer.

-1

130. It is awkward always to use parentheses to indicate sums such as $(-3) + (+2)$. If the addition sign and the parentheses are not used, the operation of addition is understood to apply. Thus, $-3 + 2$ will be used instead of $(-3) + (+2)$ to mean that -3 and $+2$ are to be added. Their sum is____.

added; -3

131. $-9 + 6$ will mean that -9 and $+6$ are to be______. Their sum is_____.

added; -2

132. $+7 - 9$ will indicate that $+7$ and -9 are to be _________. Their sum is_____.

-4

133. Generally, if a positive number occurs first in an expression, the sign + is omitted. Thus, $8 - 12$ will indicate that $+8$ and -12 are to be added. Their sum is______.

-3; 12

134. $15 - 3$ means that 15 and____are to be added. Their sum is_____.

added

135. $-8 + 7$ means that -8 and $+7$ are to be__________. Their sum is -1.

-28; -52

136. $-24 - 28$ means that -24 and_____are to be added. Their sum is_____.

2

Think $(+8) + (-6) = 2$.

137. $8 - 6 =$____.

-4

Think $(+25) + (-29) = -4$.

138. $25 - 29 =$____.

9

139. $-7 + 16 =$____.

-27

140. $-18 - 9 =$____.

-3

141. $-8 + 5 =$____.

8

142. $-19 + 27 =$____.

Remark. Is it clear yet how we do this? If two numbers have opposite sign, when we add them we are going to obtain a sum that is the *difference* of the absolute values of the numbers. If we want to find the sum of -8 and 2, we can think of it this way: The addition involves counting from the origin, and since -8 is located farther from the origin than 2, we are going to end up on the negative side of 0. How far negative? Why just the difference between eight units and two units! That is, we will find that we are six units to the *left* of the origin. Therefore, $-8 + 2 = -6$.

-15

143. $3 - 18 =$____.

8

144. $8 - 0 =$____.

-8

145. $-8 + 0 =$____.

-33

146. $-8 - 25 =$____.

24

147. $-3 + 27 =$____.

0

148. $-15 + 15 =$____.

6

149. $3 - 5 + 8 =$ ____.

$(3 - 5) + 8 = -2 + 8 = 6.$

12

150. $8 - 3 + 7 =$ ____.

-6

151. $-1 - 2 - 3 =$ ____.

-6

152. $-8 + 9 - 7 =$ ____.

-20

153. $13 - 25 - 8 =$ ____.

0

154. $0 - 7 + 7 =$ ____.

0

155. $8 - 16 + 8 =$ ____.

-8

156. $-8 + 3 + 4 - 7 =$ ____.

2

157. $2 + 3 - 8 + 5 =$ ____.*

Remark. We have now discussed the sum of two integers as a number associated with a point on a line graph arrived at by a successive counting process. We have assumed that the commutative and associative laws of addition are valid for integers. You are now able to add any number of integers. We shall next consider the meaning of the "difference" of two integers.

* See Exercise 1, page 81, for additional practice.

−3

158. The expression 5 − 3 can be looked at two ways, as the *sum* of 5 and −3, or as the *difference* of 5 and 3. That is, 3 subtracted from 5 is the same number as _____ added to 5.

difference; sum

159. 15 − 4 can be looked at two ways, as the _______________ of 15 and 4 or as the _____ of 15 and −4.

positive

160. Recall that parentheses can be used to help clarify the meaning of an expression. Thus, 7 + 3 can be written (7) + (3), where it is clear that positive seven is to be added to ___________ three.

negative

161. Parentheses can be used two ways to help clarify the meaning of 7 − 2. For example, (7) − (2) means the difference of positive seven and positive two, while (7) + (−2) means the sum of positive seven and _______________ two.

negative

162. Subtracting an integer b from an integer a, $a - b$, is the same as *adding* the _______________ of b to a, $a + (-b)$.

6

163. The difference (9) − (3) can be written as the sum (9) + (−3), because both expressions represent the same number, namely, _____.

−5

164. The difference of two *natural* numbers, $a - b$, where a is greater than b is the same as the sum of the integers a and the negative of b; that is, $(a) - (b)$ is the same as $(a) + (-b)$. The difference (7) − (5) is the same as the sum (7) + (____).

2

165. $(7) - (5) = (7) + (-5) =$____.

-2

166. $(5) - (2) = (5) + ($____$)$.

3

167. $(5) - (2) = (5) + (-2) =$____.

-2

In this case the difference is a negative number.

168. Consistent with the definition of $a - b$, where a and b are natural numbers and a is greater than b, the difference of two *integers*, $a - b$, is defined to be $a + (-b)$ whether a is greater than, equal to, or less than b. Thus, $(2) - (5) = (2) + (-5) = -3$ and $(7) - (9) = (7) + (-9) =$____.

-8

169. $(2) - (8) = 2 + ($____$)$.

-6

170. $(2) - (8) = 2 + (-8) =$____.

-16

171. $(9) - (16) = (9) + ($____$)$.

-7

172. $(9) - (16) = (9) + (-16) =$____.

$-d$

173. Recall that the difference a − b is equal to the sum $a + (-b)$, that is, the sum of a and the negative of b. $c - d$ equals $c + ($____$)$.

negative

174. $u - v$ equals u plus the____________of v.

8

175. Since the negative of -2 is 2, the difference $(7) - (-2) = (7) + (2)$. Since the negative of -8 is 8, the difference $(12) - (-8) = (12) + (\_\_\_\_)$.

5

176. The negative of -5 is_____.

5; 16

177. $(11) - (-5) = (11) + (____) =$______.

4

178. The negative of -4 is_____.

4; 19

179. $(15) - (-4) = (15) + (\quad) =$______.

Remark. Is the meaning of what we are doing here clear? That is, is it clear that we subtract y from x by simply adding $-y$ to x? If y is negative, $-y$ is positive, and we sometimes find ourselves adding positive numbers while engaged in subtracting. In the preceding frame, for instance, we subtracted -4 from 15 by adding $+4$ to 15.

21; 48

180. $(27) - (-21) = (27) + (____) =$_____.

9

181. $7 - (-2) =$_____.

$(7) + (2) = 9.$

17

182. $6 - (-11) =$_____.

negative

183. $2 - (-3)$ means to add the______________of -3 to 2.

3

184. The negative of -3 is 3, so that $2-(-3)$ means to add_____ to 2.

11

185. $4-(-7)=$_____.

-5

186. $-7-(-2)=$_____.

Remark. We are spending a lot of time on expressions such as $3-(-2)$ and $-2-(3)$ because an ability to simplify such expressions is essential in everything that follows. It is very helpful if subtraction is always thought of in terms of addition, that is, if subtracting x is thought of in terms of adding $-x$.

We have used parentheses to help clarify the meaning of expressions but it is common practice to omit the use of parentheses in such expressions altogether.

-7

187. It is common practice to write the difference $(a)-(b)$ as simply $a-b$. Since $(a)-(b)$ is the same as $(a)+(-b)$, the expression $a-b$ can be thought of as the sum of a and the negative of b. Thus, $3-7$ can be thought of as $3+($_____$)$.

negative

188. The difference $r-s$ is the same as the sum of r and the_____________ of s.

-2; -5

189. The expression $-2-5$ should be thought of as the *sum* of______ and______.

-19

190. $-2-5=-7$, because the sum of -2 and -5 is -7. Similarly, $-3-16=$______.

7

191. $18-11=$______.

-8 192. $-5 - 3 =$_____.

-3 193. $6 - 9 =$_____.

-6 194. $4 - 10 =$_____.

-12 195. $-4 - 8 =$_____.

Remark. Because the idea is so important, we emphasize again that $a - b$ can be looked at as $(a) - (b)$ or as $(a) + (-b)$, the meaning of the two expressions being the same. Thus, to subtract the number b from the number a, the negative of b can be added to a.

-18 196. $7 - 18$ means either $(7) - (18)$ or $(7) + ($_____$)$.

-3 197. $8 - 3$ means either the difference of 8 and 3 or the sum of 8 and_____.

5 198. $8 - 3$ is the same as $8 + (-3)$; both are equal to _____.

-5 199. $7 - 12$ is the same as $7 + (-12)$; both are equal to_____

-8 200. $7 - 12$ is read as the sum of 7 and -12. $10 - 8$ represents the sum of 10 and_____.

$8 - 5$

Or 3.

201. Since the difference $(10) - (8)$ is the same as the sum $(10) + (-8)$, either of these can be written as $10 - 8$. Since the difference $(8) - (5)$ is the same as the sum $(8) + (-5)$, either of these can be written as ____.

$-5 + 3$

Or -2.

202. $-7 - (-2)$ can be written as $-7 + 2$. Similarly $-5 - (-3)$ can be written as ________.

-3

203. To find a value for expressions such as $-8 - (3) - (-4)$, the entire expression can be written without parentheses as a sum, $-8 - 3 + 4$, which can then be rewritten as -7. $-5 - (6) - (-8)$ means $-5 - 6 + 8$, which can be rewritten as the single numeral_____.

-4

204. $-8 - (3) - (-7)$ means $-8 - 3 + 7$, which can be rewritten as the single numeral _____.

-2

205. $3 - (-2) - (7) = 3 + 2 - 7 =$ _____.

-28

206. $-6 - (18) + (-4) = -6 - 18 - 4 =$ _____.

-9

207. To "simplify" an expression such as $-8 + (-3) - (-2)$ means to carry out the computations indicated. Thus, $-8 + (-3) - (-2)$ can be written as $-8 - 3 + 2$, whose value is_____.

-4

208. Simplify: $-6 + (-8) - (-10)$.

$$-6 + (-8) - (-10) = -6 - 8 + 10 = -4.$$

-20

$-3+(-7)-(10) = -3-7-10$
$= -20.$

209. Simplify: $-3 + (-7) - (10)$.

4

210. Simplify: $8 - (-8) + (-12)$.

29

211. Simplify: $7 - (-7) + 15$.

11

212. Simplify: $8 - (3) - (-6)$.*

Remark. We are now going to turn our attention to polynomials. The replacement set of the variables in the polynomials will be understood to be the set of integers. That is, any variable in the polynomial will represent an integer. We shall begin our work with polynomials by reviewing the meaning of some words that we will be using.

zzz

213. Recall that a power is a product of a given number of identical factors. Thus, $x^2 = xx$ and $y^4 = yyyy$. z^3 can be written in factored form as ______.

$3 \cdot 3 \cdot 3 \cdot 3$

214. 3^4 can be written in factored form as __________.

$xxxxx$

215. x^5 can be written in factored form as __________.

product

216. Recall that an expression such as xy or $3x$ represents a product. xy represents the product of x and y; $3x$ represents the__________of 3 and x.

product

217. $4b$ represents the__________of 4 and b.

* See Exercise 2, page 81, for additional practice.

6; x; y

Any order will do.

218. $6xy$ represents the product of_____and_____and _____.

term

219. A product such as $6xy$ is commonly called a term. $7yz$ is a ______.

coefficient

220. Recall that in a term such as $3x^2y$, the symbol 3 is called the numerical coefficient or simply the coefficient of the term. In the term $4y$, 4 is called the _______________ of the term.

-9

221. The coefficient of the term $-9xy^2$ is_____.

1

222. If no numeral appears before a term, for example, xy, the coefficient 1 is understood. The coefficient of the term yz is_____.

1

223. The coefficient of the term x^2y is_____.

-1

224. Since (as we shall see later) $-x = (-1)(x)$, the term $-x$ can be viewed as having a coefficient of -1. The coefficient of the term $-y^2z$ is_____.

coefficient

225. The________________ of the term $-rs$ is -1.

like

226. Recall that like terms are terms whose variable parts are identical. Thus, because the variable factors in the terms are the same, namely, x, $3x$, $-x$, and $17x$ are_________terms.

$8x$

227. We shall assume that the distributive law in the form $ba + ca = (b + c)a$ is valid for integers as well as natural numbers. Thus, like terms in algebraic expressions can be combined by adding the coefficients of the terms. For example $7y + 2y = (7+2)y = 9y$ and $8z + z = (8+1)z = 9z$. Similarly $3x + 5x = (3+5)x =$ ________.

$4y$

$3y + y = (3 + 1)y = 4y$.

228. $3y + y =$_____.

$8a$

229. $7a + a =$_____.

$5r + 2s$

Or $2s + 5r$ if you like.

230. If an expression contains some like terms and some unlike terms, the expression can be simplified by combining such like terms as exist. For example, $3x + 2y + 2x = 5x + 2y$ and $7a + b + 3a = 10a + b$. $2r + 2s + 3r =$___________.

$3s + 6t$

231. $3s + 2t + 4t =$___________.

$18x + 5y$

232. $10x + 5y + 8x =$___________.

$3z$

233. The process of simplifying expressions containing like terms is unchanged if the coefficients of some or all of the terms are negative. Thus,
$5x - 2x = (5 - 2)x = 3x$ and
$-11y + 2y = (-11 + 2)y = -9y$.
$-2z + 5z = (-2 + 5)z =$_____.

x

$3x - 2x = (3 - 2)x = x$.

234. $3x - 2x =$_____.

$-2z$ | **235.** $7z - 9z =$ _____.

$9r$ | **236.** $10r - r =$ _____.

$4x + 2y$ | **237.** $14x + 2y - 10x =$ ____________.

$3m - n$ | **238.** $7n - 8n + 3m =$ ___________.

$2x^2 - x$ | **239.** Since x^2 and x are not like terms, they cannot be combined. Thus, $2x^2 + x + 3x^2 = 5x^2 + x$.
$5x^2 - x - 3x^2 =$ ____________.

$8y^2 - 2y$ | **240.** $5y^2 - 2y + 3y^2 =$ ____________.

$-4z^2 - 2z$ | **241.** $-7z^2 + 3z^2 - 2z =$ ______________.

$8rs + 3t$ | **242.** $10rs - 2rs + 3t =$ ____________.

Remark. Is the idea involved here clear? If we have an expression that contains some like terms and some unlike terms, we can simplify the expression by combining such like terms as there are, and then indicating the addition of any unlike terms that remain.

$11r + 3rs$ | **243.** Since rs and r are not like terms, they cannot be combined. Thus, $3rs + 2r - 2rs = rs + 2r$.
$7rs + 11r - 4rs =$ ____________.

$-3x - 2xy$ | **244.** $4xy - 3x - 6xy =$ ______________.

$2y - 8yz$

245. $-5yz + 2y - 3yz =$ ____________.

$y^3 - y^2 - y$

246. $y^3 - 2y^2 + y^2 - y =$ ________________.

$-2y^2$

247. $2x^2 - 3y^2 - 2x^2 + y^2 =$ ________.

$2c^2$

248. $4c^2 + 7c - 2c^2 - 7c =$ ______.

$ab - c$

249. $3ab - 4c + 3c - 2ab =$ __________.

$9m^2 - 5m + 1$

250. $7m^2 - 2m + 3 + 2m^2 - 3m - 2 =$
_____________________.

0

251. $5h - 3k^2 + 2hk - 5h + 3k^2 - 2hk =$ ______.

$3a + 2b$

252. The expression $(a + b) + (2a + b)$ asserts that $a + b$ is to be added to $2a + b$. This expression can be written without parentheses by simply writing $a + b + 2a + b$. If $a + b + 2a + b$ is simplified, the result is_____________.

$x^2 - 2x + 5y - 3$

253. If an expression is included in parentheses, and if the expression is preceded by a positive sign, the expression can be rewritten without the parentheses by simply indicating the addition of each term enclosed by the parentheses. For example, $(3k^2-2k)+(-3h-2)$ can be rewritten $3k^2 - 2k - 3h - 2$. Similarly, $(x^2 - 2x) + (5y - 3)$ can be rewritten________________.

$5r^2 - 2s^2 - r - 2s$

254. $(5r^2 - 2s^2) + (-r - 2s)$ can be rewritten without parentheses as____________________________.

Answer	Frame
$a + b$	**255.** $(2a - b) + (-a + 2b) = 2a - b - a + 2b =$__________.
$6x - y - z$	**256.** $(3x - y) + (3x - z) =$______________.
$2a^2 + 2a - 6$	**257.** $(a^2 + 2a - 3) + (a^2 - 3) =$________________.
$-x^2 - 4x - 1$	**258.** $(-2x^2 - 4x + 2) + (x^2 - 3) =$________________.
$-2y^2 + 3$	**259.** $(2y^3 - 3y^2 + 7) + (-2y^3 + y^2 - 4) =$______________.
$4a^2 - 4a + 3$	**260.** $(a^2 - 2a + 1) + (3a^2 - 2a + 2) =$________________.*
$-y$	**261.** Recall that the expression $a - b$ means the sum of a and the negative of b. Thus, $a - b = a + (-b)$. $2x - y = 2x +$ (____).
$-3y$	**262.** $x - 3y = x +$ (______).
$-y - 3$ Each term of $(y + 3)$ has been replaced by its negative.	**263.** The expression $a - (b + 2)$ means the sum of a and the negative of $(b + 2)$ or $a + (-b - 2)$, where each term within the parentheses has been replaced by its negative. $x - (y + 3) = x +$ (________).
$-2x - y$	**264.** $5 - (2x + y) = 5 +$ (______________).
$-x^2 - x$	**265.** $7 - (x^2 + x) = 7 +$ (______________).

* See Exercise 3, page 82, for additional practice.

$-x^2 + 5$

266. $y - (x^2 - 2) = y + (-x^2 + 2)$ because the negative of x^2 is $-x^2$ and the negative of -2 is 2.
$x - (x^2 - 5) = x + (\_\_\_\_\_\_\_\_\_\_)$.

$z - 2$

267. $2z^2 - (-z + 2) = 2z^2 + (\_\_\_\_\_\_\_)$.

$y^2 + 3y$

268. $y^3 - (-y^2 - 3y) = y^3 + (\_\_\_\_\_\_\_\_\_\_)$.

Remark. Is the idea clear? An expression such as $-(x + y)$ can be replaced with the negative of $x + y$ which is $-x - y$, where no parentheses are involved. In other words, in expressions involving only addition, parentheses preceded by a negative sign may be omitted, providing each term inside the parentheses is replaced by its negative.

$3x^2 - x - 2$

269. $x - (3z + y) = x - 3z - y$. Similarly,
$3x^2 - (x + 2) = \_\_\_\_\_\_\_\_\_\_\_\_\_\_$.

$2x - y + z$

270. $2x - (y - z) = \_\_\_\_\_\_\_\_\_\_\_\_\_\_$.

$3x - 3$

271. $x - (3 - 2x) = x - 3 + 2x = \_\_\_\_\_\_\_\_\_\_$.

$-2x - y$

272. $x - (3x + y) = x - 3x - y = \_\_\_\_\_\_\_\_\_\_$.

y

273. $2x - (-y + 2x) = \_\_\_\_$.

$4a - 2b$

274. $3a - (2b - a) = \_\_\_\_\_\_\_\_\_\_$.

$2x^2 - y^2$

275. $x^2 - (y^2 - x^2) = \_\_\_\_\_\_\_\_\_\_\_\_$.

$4y - 1$

276. $(3y + 2) - (3 - y) =$____________.

$-x^2 + 4x - 4$

277. $(3x^2 + 2x - 1) - (4x^2 - 2x + 3) =$ _________________.

0

278. $(2p^2 - 3p + 1) - (2p^2 - 3p + 1) =$____.

$-3y$

279. $(x - 2y) - (2x + y) + x =$_______.

$-2a$

280. $(2a - b) - (a - 2b) - (3a + b) =$______.

Remark. We have seen that when we have an expression containing parentheses and when the parentheses are preceded by a *negative* sign, we can rewrite the expression without parentheses providing we replace each term in the original parentheses by its negative. If an expression contains parentheses, and if the parentheses are preceded by a *positive* sign the expression can be rewritten without parentheses by simply "dropping" the parentheses. The next few frames provide you with an opportunity to work with both types of such expressions.

$z^2 + z + 3$

281. $z^2 + (z + 3) =$_________________.

$x^2 - x + 3$

282. $x^2 - (x - 3) =$_________________.

$d - 4g$

Before combining like terms, the expression appears as $3d - 2g - g - 2d - g$.

283. $(3d - 2g) - (g + 2d) - g =$_____________.

$a + 3$

284. $(a^2 - 2a + 2) + (-a^2 + 3a + 1) =$___________.

$2x + y - 3z$

285. $(x + 2y - z) - (2z - x + y) =$ ________________.

0

286. $(3x - 2y + z) + (2y - z - 3x) =$ ____.

$8a - 2b + 4$

287. $(4a - b + 2) - (b - 4a - 2) =$ ________________.

$-x - y + 4$

288. $(x - 2y + 1) - (-y + 2x - 3) =$ ________________.*

Remark. We are now ready to look at the products of integers, but first, let us recall what is meant by the product of two natural numbers or, as we now call them, two positive integers.

7 + 7 + 7

289. In arithmetic, the product of two natural numbers such as (3)(4) can be viewed as the sum of three 4's. That is, (3)(4) = 4 + 4 + 4. Similarly, (3)(7) = ___ + ___ + ___.

7 + 7

290. (6)(7) = 7 + 7 + 7 + 7 + ____ + ____ = 42.

35

291. Having learned the multiplication tables, (6)(7) can be written directly as 42. Similarly, (7)(5) can be written directly as______.

Remark. Now we will see what effect negative factors are going to have on products.

0

292. Recall that 4 + (−4) = 0. 3 + (−3) =____.

−6

293. If 6 + (x) = 0, then x represents the number____.

−14

Because (14) + (−14) = 0.

294. If (2)(4) + (x) = 0, then x represents −8, and if (2)(7) + (x) = 0, then x represents______.

* See Exercise 4, page 82, for additional practice.

-14

295. If $(3)(5) + ab = 0$, then $ab = -15$, and if $(2)(7) + (ab) = 0$, then $ab =$ _____.

-35

296. If $(5)(7) + (ab) = 0$, then $ab =$ _____.

Remark. The idea being developed here is just this: If the sum of two numbers is 0, and if both numbers are not 0, then each of the numbers is the negative of the other.

negative

297. If $(5)(7) + (ab) = 0$, then ab is the __________ of $(5)(7)$.

negative

298. If $(4)(3) + (ab) = 0$, then ab is the __________ of $(4)(3)$.

distributive

299. Recall that the distributive law as given for natural numbers is written $a(b + c) = ab + ac$. Thus, $3(2 + 5) = 3(2) + 3(5)$ is an application of the __________ law.

-3

300. Recall that we assumed that the distributive law is valid for integers as well as natural numbers. Thus, by the distributive law, $4[3 + (-3)] = 4(3) + 4(\_\_\_)$.

0

301. $3 + (-3) = 0$. Therefore, since $4(0) = 0$, the product $4[3 + (-3)]$ is equal to _____.

0

302. By the distributive law, $4[3 + (-3)] = 4(3) + 4(-3)$. Since $4[3 + (-3)]$ equals $4(0)$ or 0, then the expression $4(3) + 4(-3)$ must also equal _____.

negative

303. 4(3) + 4(−3) will equal 0 only if 4(−3) is the ____________ of 4(3).

−12

304. If 4(−3) is the negative of 4(3), then, because 4(3) is 12, 4(−3) must be ______.

−15

305. As in the previous frame, it can be shown that 2(−3) = −(2 · 3) or −6. Similarly, 3(−5) = −(3)(5) or _____.

−18

306. In general, $a(-b) = -ab$. Thus, 6(−3) = _____.

−8

307. 4(−2) = _____.

−66

308. Recall that the commutative law of multiplication for natural numbers says that $ab = ba$. It will be assumed that the multiplication of integers is also commutative. Thus, (−6)(5) = (5)(−6) = −30, (−10)(7) = (7)(−10) = −70, and (−6)(11) = (11)(−6) = ______.

−42

309. (−6)(7) = _____.

−42

310. (7)(−6) = ______.

−12

311. (−6)(2) = _____.

−40

312. (8)(−5) = _____.

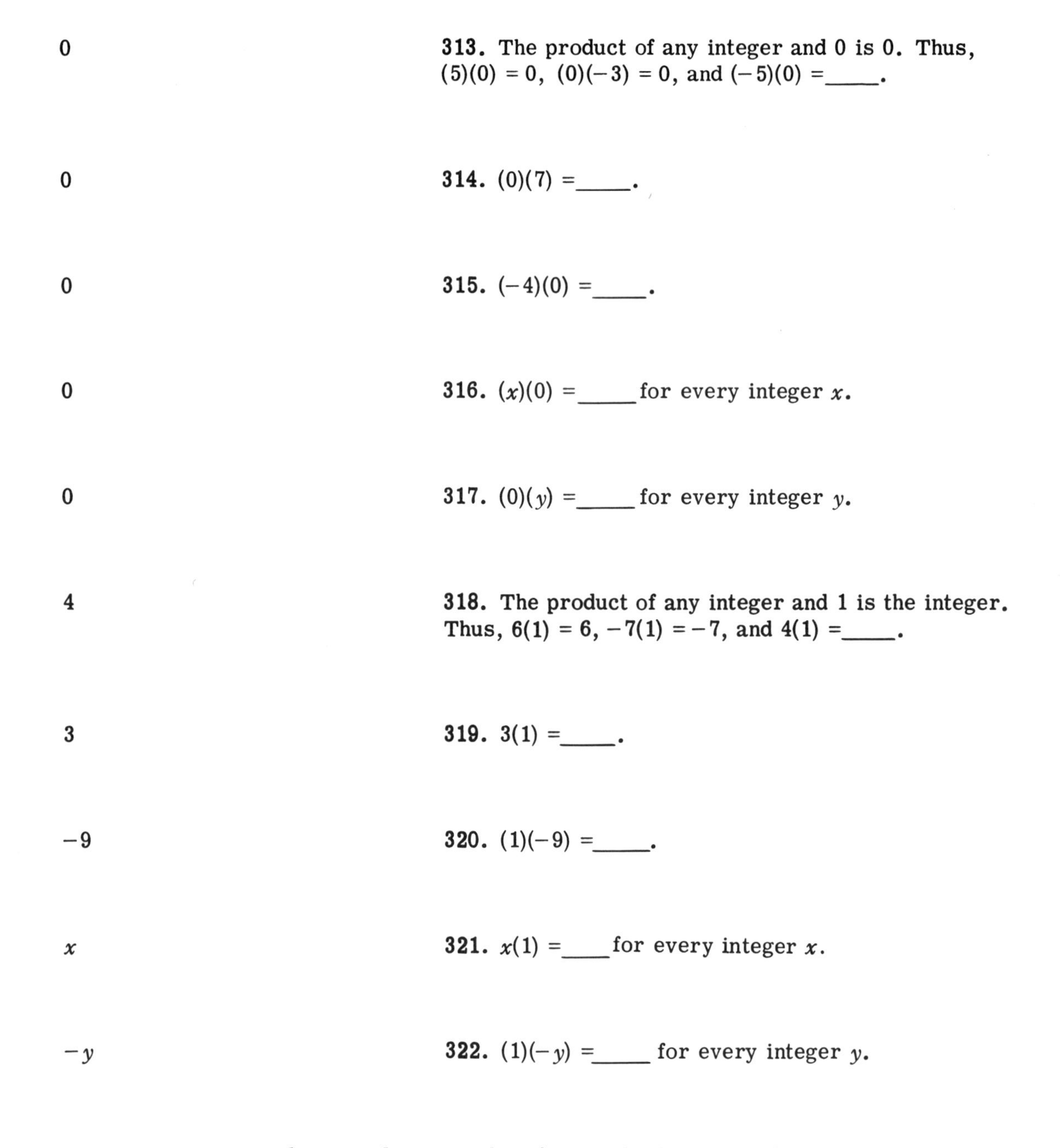

0

313. The product of any integer and 0 is 0. Thus, $(5)(0) = 0$, $(0)(-3) = 0$, and $(-5)(0) =$ ____.

0

314. $(0)(7) =$ ____.

0

315. $(-4)(0) =$ ____.

0

316. $(x)(0) =$ ____ for every integer x.

0

317. $(0)(y) =$ ____ for every integer y.

4

318. The product of any integer and 1 is the integer. Thus, $6(1) = 6$, $-7(1) = -7$, and $4(1) =$ ____.

3

319. $3(1) =$ ____.

-9

320. $(1)(-9) =$ ____.

x

321. $x(1) =$ ____ for every integer x.

$-y$

322. $(1)(-y) =$ ____ for every integer y.

Remark. Is it clear now that if we multiply two numbers that are opposite in sign, the product is a negative number ? We shall next consider the result of multiplying two negative numbers.

0

323. Recall that $(-3) + (+3) = 0$; $(-5) + (+5) =$_____.

15

324. If $-7 + (x) = 0$, then x must represent an integer that adds to -7 to yield 0. Thus, x must equal 7. If $-15 + (x) = 0$, then x must equal_____.

12

325. If $(-6) + (x) = 0$, then x equals 6. If $(-12) + (x) = 0$, then x equals______.

42

326. If $(-7)(5) + ab = 0$, then ab must represent an integer that adds to -35 to yield 0. Thus, ab must equal 35. If $(-6)(7) + ab = 0$, then ab must equal_____.

15

327. If $(-3)(4) + (ab) = 0$, then ab equals 12. If $(-3)(5) + (ab) = 0$, then $ab =$______.

6

328. If $(-2)(3) + (ab) = 0$, then ab equals_____.

0

329. By the distributive law, $-3[5 + (-5)] = (-3)(5) + (-3)(-5)$. Since $-3[5 + (-5)]$ also equals $-3(0)$ or 0, the expression $(-3)(5) + (-3)(-5) =$_____.

20

330. If $(-3)(5) + (-3)(-5) = 0$, the product $(-3)(-5)$ must be the negative of the product $(-3)(5)$ because their sum is zero. Since $(-3)(5) = -15$, $(-3)(-5)$ equals 15. Since $(-4)(5) = -20$, $(-4)(-5)$ must be the negative of -20. $(-4)(-5)$ equals______.

28

331. As in the previous frame, it can be shown that $(-2)(-3) = 6$, $(-5)(-7) = 35$, and, indeed, that the product of any two negative integers is a positive integer. Thus, $(-7)(-4) =$______.

24 | 332. $(-3)(-8) =$ _____.

7 | 333. $(-1)(-7) =$ _____.

Remark. By now it should be clear that if we multiply two integers of like sign, the product is a positive integer. That is, the product of two positive integers is a positive integer and the product of two negative integers is also a positive integer. Furthermore, if we multiply two integers with opposite signs, the product is a negative integer. We shall reinforce this idea with some more practice.

-28 | 334. $(-7)(4) =$ _____.

-30 | 335. $(-6)(5) =$ _____.

30 | 336. $(-6)(-5) =$ _____.

70 | 337. $(-7)(-10) =$ _____.

-18 | 338. $(-9)(2) =$ _____.

0 | 339. $(0)(-7) =$ _____.

-63 | 340. $(7)(-9) =$ _____.

17 | 341. $(-17)(-1) =$ _____.

-25

342. $(-1)(25) =$ ______.

24

343. Recall that the associative law of multiplication for natural numbers says that if a, b, and c are natural numbers, $(ab)c = a(bc)$. Thus, $(2)(3)(4)$ means $[(2)(3)](4)$ or $(2)[(3)(4)]$. Either product equals the same number, ______.

24

344. Assuming that the associative law of multiplication is also true for integers, the product of three or more integers, such as $(-2)(-3)(4)$ will mean $[(-2)(-3)](4)$ or $(-2)[(-3)(4)]$. Either product equals ______.

30

345. $(-2)(3)(-5)$ can be computed by multiplying -2 by 3 to obtain -6 and then multiplying -6 by -5 to obtain 30. $(-2)(3)(-5)$ can also be computed by multiplying 3 by -5 to obtain -15, and then multiplying -15 by -2 to obtain ______.

-24

$(6)(-4) = -24$ or $(2)(-12) = -24$.

346. $(2)(3)(-4) =$ ______.

15

347. $(-5)(-1)(3) =$ ______.

36

348. $(2)(-3)(-6) =$ ______.

-15

349. $(-1)(-3)(-5) =$ ______.

-1

350. $(-1)(-1)(-1) =$ ______.

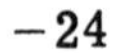

−24

351. $(2)(4)(-1)(3) =$ ______.

−6

352. $(-3)(-2)(-1)(1) =$ ______.

0

353. If 0 is a factor of a product, then the product must be 0. Thus, $(3)(0) = 0$, $(-2)(4)(0) = 0$, and $(-4)(7)(0) =$ ____.

0

354. $(-7)(8)(5)(0) =$ ____.

Remark. Thus far we have concentrated on the multiplication of integers. This multiplication obeys the commutative and associative laws. We are going to begin now with the multiplication of terms involving variables.

five

355. Since x occurs as a factor twice in x^2 and three times in x^3, then in the product $x^2 \cdot x^3$, x must occur as a factor ________ times.

six

356. Since y occurs as a factor three times in y^3, then in the product $y^3 \cdot y^3$, y must occur as a factor ______ times.

six

If necessary, write the factors out and count them, *zzzzzz*.

357. In z^4, z occurs as a factor four times, while in z^2, z occurs as a factor twice. In the product $(z^4)(z^2)$, z occurs as a factor ______ times.

nine

358. $(b^3)(b^2)$ means that the factor b occurs five times in the product. $(b^7)(b^2)$ means that the factor b occurs ______ times in the product.

b^9

359. $(b^7)(b^2) =$ _____.

x^4

360. $(x)(x^3) =$ _____.

y^6

361. $(y)(y^5) =$ _____.

k^8

362. $(k)(k^7) =$ _____.

Remark. Think of it this way: If you want the product of two powers with like bases, all you have to do is write a power of the same base whose exponent is the *sum* of the exponents of the factors; that is, $x^m \cdot x^n = x^{m+n}$.

x^{14}

363. $x^3 \cdot x^{11} =$ _____.

y^{17}

364. $y^8 \cdot y^9 =$ _____.

x^{35}

365. $(x^{15})(x^{20}) =$ _____.

six

366. $x(x^2)(x^3)$ means that the factor x occurs _____ times in the product.

x^6

367. $(x)(x^2)(x^3) =$ _____.

y^7

368. $(y)(y^2)(y^4) =$ _____.

z^{11}

369. $(z)(z^5)(z^5) =$ _____.

a^3b

370. If two powers such as x^3 and y^4 have different bases, their product is written x^3y^4. The product of a^3 and b is written_____.

b^4c^2

371. The product of b^4 and c^2 is written_______.

monomials

372. Recall that an expression such as $3x^2y$, that consists of a single term, is called a <u>monomial</u>. $4x^2$, 3, $-2a^2b$, and $x^2y^2z^2$ are examples of__________________.

$-8y^4$

373. Consider the product $(3x)(2x^2)$. By the commutative and associative laws, the factors of this product can be arranged in the form $(3)(2)(x)(x^2)$, which is equal to $6x^3$. Similarly, $(2y)(-4y^3)$ can be written $(2)(-4)(y)(y^3)$, which is equal to______.

$10c^2d$

374. To multiply two monomials, it is necessary to multiply both their coefficients and their variable factors. Thus, $(3x^2)(2xy) = 6x^3y$ and $(-3a)(4a^2b) = -12a^3b$. Similarly, $(5c)(2cd) =$________.

$6xy$

Of course, $6yx$ or any other arrangement of the factors is also correct. In our answers, we will arrange the variables in alphabetical order.

375. $(3x)(2y) =$______.

$15ax$

376. $(5x)(3a) =$________.

$-21x^2y$

377. $(3x^2)(-7y) =$___________.

$-10xy^2$

378. $(-5x)(2y^2) =$ ________.

$3x^3$

379. $(-3x^2)(-x) =$ ______.

$5y^4$

380. $(5y^3)(y) =$ ______.

$7a^3$

381. $(-7a^2)(-a) =$ ______.

$-5z^6$

382. $(5z^5)(-z) =$ ______.

$2y^5$

383. $(-2y^3)(-y^2) =$ ______.

$-35x^3yz$

384. $(-5x^2z)(7xy) =$ ____________.

$24r^2s^4t$

385. $(3r^2s^2)(8s^2t) =$ ____________.

$-4x^4y$

386. $(2x)(x^2)(-2xy) =$ ________.

$-x^3$

387. $(-x)(-x)(-x) =$ ______.

y^4

388. $(-y)(-y)(-y)(-y) =$ ____.

$-2x^2$

389. $-2(x)(x) =$ ______.

$2x^2$

390. $(-2)(-x)(x) =$ ______.

$-2x^2$

391. $(-2)(-x)(-x) =$________.

$-6x^3y$

392. $-3(2x^2)(xy) =$_________.

$-12a^2b$

393. $-2(3a^2)(2b) =$____________.

$36x^2y^3z^2$

394. $(3xy)(2yz)(6xyz) =$______________.

$-21r^4s^4$

395. $(-3rs)(-s^2)(-7r^3s) =$_____________.*

Remark. The last operation with integers we will consider in this unit is that of division. As you will see, division is closely related to multiplication in that division is defined in terms of multiplication. We shall first consider the division of one integer by another, and then go directly from this into the division of one monomial by another.

quotient

396. Recall that the quotient of two expressions is generally written in the form of a fraction. Thus, $\frac{x}{y}$ denotes the______________of x divided by y.

3; z

397. $\frac{3}{z}$ denotes the quotient of____ divided by____.

5

398. The quotient $\frac{b}{a}$ is defined to be the number c such that $ac = b$. For example, the quotient $\frac{6}{2}$ is 3 because $(2)(3) = 6$. The quotient $\frac{15}{3}$ is 5 because $(3)($___$)$ is 15.

* See Exercise 5, page 83, for additional practice.

9

399. The quotient $\frac{18}{2}$ is____because $(2)(9) = 18$.

-3

400. The quotient $\frac{-10}{2}$ is -5 because $(2)(-5) = -10$. The quotient $\frac{-12}{4}$ is -3 because $(4)(\_\_\_) = -12$.

-8

401. The quotient $\frac{24}{-3}$ is_____because $(-3)(-8) = 24$.

6

402. The quotient $\frac{-36}{-6}$ is_____because $(-6)(6) = -36$.

6

Because $(-2)(6) = -12$.

403. $\frac{-12}{-2} =$_____.

-6

Because $(2)(-6) = -12$.

404. $\frac{-12}{2} =$_____.

negative

405. It is evident that the quotient of any two numbers of like sign must carry a positive sign, while the quotient of two numbers of opposite sign must carry a negative sign. Thus, if a and b are positive, the quotients $\frac{a}{b}$ and $\frac{-a}{-b}$ must both be positive, while the quotients $\frac{-a}{b}$ and $\frac{a}{-b}$ must both be______________.

2

406. The quotient $\frac{-16}{-8}$ is the number_____, which is positive.

5

407. The quotient $\frac{15}{3}$ is the number ____, which is positive.

-5

408. The quotient $\frac{-50}{10}$ is the number____, which is negative.

-8

409. The quotient $\frac{40}{-5}$ is the number____, which is negative.

9

410. $\frac{27}{3}$ =____.

6

411. $\frac{-30}{-5}$ =____.

-2

412. $\frac{-24}{12}$ =____.

-12

413. $\frac{24}{-2}$ =____.

0

414. The quotient $\frac{a}{b}$ is defined to be the number c such that $bc = a$. Therefore, the quotient $\frac{0}{b}$, where b does not equal zero, always equals 0, because $(0)(b)$ always equals 0. Thus, $\frac{0}{4}$ =____.

0

Because (2) (0) = 0.

415. $\frac{0}{2}$ =____.

0

416. $\frac{0}{-7}$ = ____.

positive

417. The quotient of two negative numbers is a ____________ number.

do

418. $\frac{-10}{-2}$ and $\frac{10}{2}$ (do/do not) represent the same number.

negative

419. The quotient of two numbers of opposite sign is a ____________ number.

do

420. $\frac{-6}{2}$ and $\frac{6}{-2}$ (do/do not) represent the same number.

1

421. The quotient of any positive number divided by itself is the number 1. For example, $\frac{4}{4} = 1$, $\frac{7}{7} = 1$, and $\frac{5}{5}$ = ____.

1

422. The quotient of any negative number divided by itself is also the number 1. For example, $\frac{-3}{-3} = 1$, $\frac{-7}{-7} = 1$, and $\frac{-9}{-9}$ = ____.

1

423. $\frac{-6}{-6}$ = ____.

undefined

424. The quotient $\frac{a}{b}$ is defined to be the number c such that $bc = a$. The quotient $\frac{3}{0}$, then, must be the number that multiplies 0 to give 3. But there is no such number, because 0 times any number is 0. Therefore, $\frac{3}{0}$ is said to be undefined. $\frac{4}{0}$ is also ____________.

meaningless

425. Since the quotient of any number divided by 0 is undefined, division by 0 is said to be meaningless. Thus, $\frac{8}{0}$ is a ________________ expression.

undefined

Or meaningless, but undefined is preferred.

426. $\frac{0}{0}$ is also undefined, because the quotient can be interpreted to be any number. Thus, it might be claimed that $\frac{0}{0} = 7$ because $(0)(7) = 0$, or that $\frac{0}{0} = 4$ because $(0)(4) = 0$. This claim could also be made for any number a, because $(0)(a) = 0$. Therefore, the quotient $\frac{0}{0}$ is said to be __________ .

undefined

427. $\frac{5}{0}$ is ______________.

undefined

428. $\frac{-3}{0}$ is ______________.

0

429. $\frac{a}{b}$ is undefined if a represents any number and b represents the number ____.

Remark. It is important to remember that we cannot give any meaning to a fraction whose denominator is zero. However, it is equally important to remember that a fraction whose numerator is zero but whose denominator is not zero has a perfectly respectable value, namely, zero. Thus, while $\frac{3}{0}$ or $\frac{0}{0}$ is undefined, $\frac{0}{4} = 0$.

There is another way in which we can look at the division process, a way that is very important because it will help us when we begin to divide expressions containing variables. Give the next sequence of frames very close attention, because you will need to understand the process they develop.

$\frac{4}{4}$

Don't consider your answer wrong if you wrote 1 for the quotient $\frac{4}{4}$. After all, $\frac{4}{4}$ is equal to 1.

430. The quotient $\frac{ab}{b}$ can be interpreted to mean $a\left(\frac{b}{b}\right)$. Thus, $\frac{2 \cdot 3}{3}$ can be written $2\left(\frac{3}{3}\right)$, $\frac{5 \cdot 7}{7} = 5\left(\frac{7}{7}\right)$, and $\frac{3 \cdot 4}{4} = 3\left(\underline{\quad}\right)$.

6

431. $\frac{6 \cdot 2}{2} = 6\left(\frac{2}{2}\right) = \underline{\quad}(1)$.

1

432. $\frac{3 \cdot 4}{3} = \left(\frac{3}{3}\right)4 = (\underline{\quad})4$.

$\frac{6}{6}$

433. $\frac{6 \cdot 15}{6} = \left(\underline{\quad}\right)15 = (1)15$.

8; 8

434. $\frac{7 \cdot 8}{7} = \left(\frac{7}{7}\right)\underline{\quad} = (1)\underline{\quad}$.

6

435. The quotient $\frac{24}{3}$ can be written $\frac{(8)(3)}{3}$ because $24 = (8)(3)$. The quotient $\frac{18}{6}$ can be written $\frac{(3)(\_\_\_)}{6}$ because $18 = (3)(6)$.

1

436. $\frac{(3)(6)}{6} = 3\left(\frac{6}{6}\right) = 3(\_\_\_) = 3$.

4

437. The quotient $\left(\frac{8}{2}\right)$ can be written $\frac{(4)(2)}{2}$ because $8 = (\_\_\_)(2)$.

$\frac{2}{2}$

438. $\frac{(4)(2)}{2} = 4\left(\frac{\quad}{\_\_\_}\right) = 4(1) = 4$.

Remark. Is it clear that what we are doing here is factoring the numerator of a fraction so that one of the factors is equal to the denominator? If we can factor a numerator this way, then we can simplify a quotient.

5

439. The quotient $\frac{15}{3}$ can be written $(\_\_\_)\left(\frac{3}{3}\right)$.

1

440. $5\left(\frac{3}{3}\right) = 5(\_\_\_) = 5$.

3

441. The quotient $\frac{24}{8}$ can be written $3\left(\frac{8}{8}\right)$, which is equal to 3(1) or____.

$\frac{2}{2}$

Or 1 is also right.

442. The quotient $\frac{4y}{2}$ can be written $\frac{(2y)(2)}{2}$, which is equal to $(2y)\left(\_\_\right)$.

$2y$

443. $(2y)\left(\frac{2}{2}\right) = (2y)(1) = \_\_\_\_$.

$3y$

444. $\frac{12y}{4} = \frac{(3y)(4)}{4} = (3y)\left(\frac{4}{4}\right) = \_\_\_\_$.

$2y$

445. If one monomial is exactly divisible by another, the quotient can be computed by writing the dividend (numerator) as the product of two factors, one of which is the divisor (denominator). Thus, to divide $4xy$ by $2x$, the quotient is first written in fractional form, $\frac{4xy}{2x}$, and then $2y\left(\frac{2x}{2x}\right)$. Since $\frac{2x}{2x}$ is equal to 1, the quotient becomes_____.

$4y$

446. $\frac{12y^2}{3y} = \frac{(4y)(3y)}{3y} = 4y\left(\frac{3y}{3y}\right) = \_\_\_\_$.

0

447. When dividing one monomial by another, it is to be understood that any variable contained in the denominator must not have a value of zero. Thus, $\frac{x^2}{x} = x\left(\frac{x}{x}\right) = x$ for all values of x except____.

0

448. $\frac{3y^2z}{3z} = y^2\left(\frac{3z}{3z}\right) = y^2$ for all values of y, and for all values of z except____.

0

449. $4a\left(\frac{3ab}{3ab}\right) = 4a$ for all values of a and b except___.

Remark. As has been shown, we cannot permit fractions to have denominators of zero. We will not pursue this idea further in this unit, but we will make an agreement now that in any fraction that contains a variable in the denominator, the variable will not assume a value that would cause the denominator to become zero.

$2xy$

450. $\frac{24x^2y}{12x} = 2xy\left(\frac{12x}{12x}\right) =$ ______.

$2x$; $2x$

451. $\frac{8x}{4} =$ ___ $\left(\frac{4}{4}\right) =$ ___.

6; 6

452. $\frac{12x}{2x} =$ ____ $\frac{2x}{2x}$ $=$ ___.

x; x

453. $\frac{x^2y}{xy} =$ ___ $\frac{xy}{xy}$ $=$ ___.

$3a$; $3a$

454. $\frac{12a^2b^2}{4ab^2} =$ ____ $\frac{4ab^2}{4ab^2} =$ ____.

Remark. From now on we will not write such things as $\frac{x^2}{x} = x\left(\frac{x}{x}\right) = x$, but will omit the step $x\left(\frac{x}{x}\right)$. This step should be accomplished mentally, and the quotient written directly from the original fraction. Thus, we will just write such things as $\frac{xy^2}{y} = xy$, $\frac{6x^2}{2x} = 3x$, and so on. Of course, if you cannot accomplish the division mentally, you can always write in the missing step, but the more you can do mentally, the better off you will be.

$2xy$

455. $\frac{6x^2y}{3x} =$ ______.

$2xy$

456. $\frac{14xy^3}{7y^2} =$ ______.

$5a$

457. $\frac{20a^2bc}{4abc} =$ ______.

ab^5

458. $\frac{a^5b^7}{a^4b^2} =$ ______.

y

459. $\frac{9x^3y}{9x^3} =$ ______.

b

460. $\frac{4a^3b^3}{4a^3b^2} =$ ______.

positive; negative

461. Recall that, when simplified, the quotient of two numbers of like sign must carry a positive sign, while the quotient of two numbers of opposite sign must carry a negative sign. Thus, when simplified, the quotients $\frac{2x^2y}{xy}$ and $\frac{-3x}{-x}$ both carry ______________ signs and the quotients $\frac{-6xy}{2y}$ and $\frac{4xy^2}{-2xy}$ both carry ____________ signs.

negative

462. When simplified, the quotient $\frac{-4ab^2}{a}$ carries a (positive/negative) sign.

positive

463. When simplified, $\frac{-x^2y^3}{-xy}$ carries a (positive/negative) sign.

$-2x$

464. $\dfrac{-6x^2}{3x} =$ _____.

$2a^2$

465. $\dfrac{-4a^3b}{-2ab} =$ _____.

Remark. Since the quotients $\dfrac{a}{b}$, $\dfrac{-a}{b}$, $\dfrac{a}{-b}$, and $\dfrac{-a}{-b}$ are always the same except for sign, in simplifying a quotient such as $\dfrac{-2x^2}{x}$, you simplify $\dfrac{2x^2}{x}$, and noting that the result is negative, write the result with a negative sign. In this example, the result will be $-2x$.

$-4y$

466. $\dfrac{24x^3yz}{-6x^3z} =$ _____.

$-2c$

467. $\dfrac{-6b^2c}{3b^2} =$ _____.

$5xy$

468. $\dfrac{5x^2y^3}{xy^2} =$ _____.

-1

469. $\dfrac{-5a^2b^2}{5a^2b^2} =$ _____.

1

470. $\dfrac{-2a^3}{-2a^3} =$ ____.

$5bc$

471. $\dfrac{15abc^3}{3ac^2} =$ _____.

$-2b^3$

472. $\frac{-4ab^3c}{2ac} =$ ______.

$-2yz$

473. $\frac{16xy^3z^3}{-8xy^2z^2} =$ ______.

$5xyz^3$

474. $\frac{-20x^3y^3z^3}{-4x^2y^2} =$ ______.*

Remark. You have now learned to add, subtract, multiply, and divide with integers, and with expressions containing variables. We shall next consider the process of simplifying expressions that involve more than one operation at a time. This will make it necessary for us to agree on the order in which we will perform the operations whenever more than one operation occurs in the same expression.

14; 14

475. In any expression involving only multiplication and division, either operation may be performed first. For example, $\frac{2 \cdot 6}{3} = \frac{12}{3} = 4$, and $\frac{2 \cdot 6}{3} = 2\left(\frac{6}{3}\right) = 2 \cdot 2 = 4$. Similarly, $\frac{6 \cdot 7}{3} = \frac{42}{3} =$ ______; or $\frac{6 \cdot 7}{3} = \left(\frac{6}{3}\right)7 = 2 \cdot 7 =$ ______.

6; 6

476. $\frac{24 \cdot 2}{8} = \frac{48}{8} =$ ____ ; or $\frac{24 \cdot 2}{8} = \left(\frac{24}{8}\right)2 = 3 \cdot 2 =$ ____.

multiplication

477. In any expression involving multiplication and addition, the multiplication operation should be performed first. For example, to simplify $3 \cdot 2 + 4$, we write $3 \cdot 2 + 4 = 6 + 4 = 10$, where the operation of __________________ has been performed before the operation of addition.

* See Exercise 6, page 84, for additional practice.

16

478. $6 \cdot 3 - 2 = 18 - 2 =$ _____.

Multiply $6 \cdot 3$ first, then add -2.

33

479. $7 \cdot 5 - 2 =$ _____.

22

480. $4 + 3 \cdot 6 =$ _____.

Multiply $3 \cdot 6$ first, then add 4.

-2

481. $6 - 2 \cdot 4 =$ _____.

29

482. $5 \cdot 7 - 2 \cdot 3 =$ _____.

First multiply $5 \cdot 7$ and $2 \cdot 3$, then add 35 and -6.

-11

483. $6 \cdot 4 - 7 \cdot 5 =$ _______.

18

484. Parentheses can be used to indicate that a sum is to be viewed as a single number. For example, $(4 + 2)3$ means that $4 + 2$ should be viewed as a single number, 6, and this number should be multiplied by 3. Thus, $(4 + 2)3 = (6)3 =$ _____.

20

485. Where possible, operations contained within parentheses should be performed first. $5(6 - 2)$ means $5(4)$ or _____.

2; 6

486. $3(7 - 5) = 3($ ___ $) =$ ____.

-8

487. $4(5 - 7) =$ _____.

30

488. $6(6 - 1) =$ ______.

9

489. $(6 - 3)3 =$ ____.

35

490. $(5 + 2)5 =$ _____.

division

491. In any expression in which both division and addition are involved, the division operation should be performed first. For example, to simplify $\frac{6}{2} + 4$, we write $\frac{6}{2} + 4 = 3 + 4 = 7$, where the operation of ____________ has been performed before the operation of addition.

-2

492. $\frac{-8}{2} + 2 =$ _____.

First divide -8 by 2, then add -4 and 2.

8

493. $6 + \frac{6}{3} =$ ____.

-2

494. $\frac{5}{5} - 3 =$ _____.

1

495. $4 - \frac{6}{2} =$ ____.

2

496. $\frac{8}{2} - \frac{12}{6} =$ ____.

3

497. $\frac{6}{2} + \frac{12}{4} - \frac{9}{3} =$____.

2

498. Since parentheses can be used to indicate that a sum is to be viewed as a single number, the expression $(8 + 4) \div 6$ means that $8 + 4$ should be viewed as the single number 12 and this number should be divided by 6. Thus, $(8 + 4) \div 6 = 12 \div 6 =$____.

2

499. In fractional form the quotient $(8 + 4) \div 6$ appears $\frac{8 + 4}{6}$, where the fraction bar serves the same purpose as parentheses, and the sum $8 + 4$ is viewed as a single number, 12. Thus, $\frac{8 + 4}{6} = \frac{12}{6} =$____.

2

500. $\frac{6 + 2}{4} = \frac{8}{4} =$____.

1

501. $\frac{6 - 3}{3} =$____.

-2

502. $\frac{4 - 18}{7} =$____.

3

503. $\frac{18 - 3}{5} =$____.

-6

You first have $\frac{12}{4} - 9$ and then $3 - 9$.

504. $\frac{6 + 6}{4} - 9 =$____.

3

505. $\frac{8-2}{3}+1=$ ____.

2

506. $\frac{4+6}{3+2}=$ ____.

You first have $\frac{10}{5}$.

2

507. $\frac{16-8}{3+1}=$ ____.

1

508. $\frac{8+16}{30-6}=$ ____.

36−1; 35

509. In any expression involving powers, the operation of multiplication associated with the power is performed first. Thus $3^2+1=9+1=10$. Similarly, $6^2-1=$ ______ $-1=$ _____.

19

510. $4^2+3=$ _____.

You first have 16 + 3.

−17

511. $8-(5)^2=$ _____.

You first have 8 − 25.

47

512. $7^2-2=$ _____.

−5

513. $4-3^2=$ ____.

17

514. $8 + (-3)^2 =$______.

Remark. It is worth noting here that there is a difference between -2^2 and $(-2)^2$. In the first case, the exponent applies only to 2, the negative sign is not involved in the squaring process, while in the second case, the exponent applies to -2 and the negative sign is involved. -2^2 can be looked at as $(-1)2^2$, or $(-1)(2)(2)$, while $(-2)^2 = (-2)(-2)$. Be careful of this distinction as you continue.

21

515. $12 + (-3)^2 =$______.

3

516. $12 - 3^2 =$_____.

7

517. $8 - (-1)^2 =$_____.

4

518. $\frac{6^2}{9} = \frac{36}{9} =$_____.

8

519. $\frac{4^2}{2} =$_____.

1

520. $\frac{(-3)^2}{9} =$_____.

3

You first have $\frac{3+9}{4}$.

521. $\frac{3 + 3^2}{4} =$_____.

3

522. $\frac{13 - (-2)^2}{3} =$_____.

3

523. $\frac{6 + 3^2}{5}$ = _____.

-1

524. $\frac{4 - (3)^2}{5}$ = ____.

4

You first have $\frac{25 - 9}{4}$.

525. $\frac{5^2 - 3^2}{2^2}$ = _____.

-1

526. $\frac{3^2 - 5^2}{4^2}$ = ______.

12

You first have $3 \cdot 4$.

527. $3 \cdot 2^2$ = ______.

54

528. $6 \cdot 3^2$ = _____.

-20

529. $-5 \cdot 2^2$ = _______.

27

530. $3 \cdot 3^2$ = ______.

44

531. $4 \cdot 3 + 2 \cdot 4^2$ = ______.

-36

532. $6 \cdot 2 - 3 \cdot 4^2$ = _______.

14

533. $2 \cdot 5^2 - 4 \cdot 3^2$ = ______.

Remark. The next sequence of frames contains a variety of expressions to simplify and involves all of the basic operations.

8

534. $\frac{4 \cdot 6}{3} =$ ____.

30

535. $5 \cdot 7 - 5 =$ ____.

-30

536. $5(2 - 8) =$ ____.

2

537. $5 - \frac{6}{2} =$ ____.

-1

538. $\frac{6 - 8}{2} =$ ____.

1

539. $\frac{5 + 10}{3} - 4 =$ ____.

19

540. $4^2 + 3 =$ ____.

5

541. $6 - (-1)^2 =$ ____.

-33

542. $3 \cdot 2^2 - 5 \cdot 3^2 =$ ____.

2

543. $\frac{4^2 + 2}{9} =$ ____.

4

544. $\frac{2(3)^2 + 2}{5} =$_____.

−11

545. $\frac{3 - 5^2}{2} =$_______.

0

546. $\frac{3^2 - 5^2}{4} + 4 =$_____.

−3

547. $(3^2 - 18) \div 3 =$______.

−4

548. $(8 - 4^2) \div (3 - 1) =$______.*

Remark. Now that we have established the order in which the basic operations are performed when more than one operation occurs in an expression, we are ready to apply this concept to find values for expressions containing variables. Careful work is required here if you are to avoid errors.

12

549. For each value of the variable, the product $3y$ represents a number. If the variable y is replaced by 2, the product becomes 3(2) or 6. If the variable y is replaced by 4 the product becomes 3(4) or______.

6

550. For each value of the variable, the quotient $\frac{x}{2}$ represents a number. If the variable x is replaced by 8, the quotient is $\frac{8}{2}$ or 4. If the variable x is replaced by 12, the quotient is $\frac{12}{2}$ or_____.

* See Exercise 7, **page 84**, for additional practice.

evaluated

551. The process of finding a value for an algebraic expression when the variables have been replaced by specified numbers is called numerical evaluation. The expression is said to be evaluated for the specified values of the variables. If the variable y in the sum $y + 3$ is replaced by -4, the sum becomes $-4 + 3$ or -1. The expression $y + 3$ is then said to have been __________ for a value of y equal to -4.

evaluated

552. If y is replaced by 9 in $\frac{2y}{3}$, the quotient becomes $\frac{2(9)}{3}$ or 6. The expression $\frac{2y}{3}$ is then said to have been ______________ for a value of y equal to 9.

3

553. The expression $x + 1$ can be evaluated for x equal to 3 by replacing x with 3 and writing ____ $+ 1$ or 4.

(−4)

554. When x is replaced by -4 in $2x + 3$, the expression becomes $2\,(\_\_\_) + 3$.

−5

555. $2(-4) + 3 =$ _____.

2

3(−1) + 5,
−3 + 5.

556. Evaluate $3y + 5$ for $y = -1$.

12

2(3) + 6,
6 + 6.

557. Evaluate $2x + 6$ for $x = 3$.

-2

558. When x is replaced by -2 in $4 - 3x$, the expression becomes $4 - 3(\_\_\_)$.

10

$4 - 3(-2)$,
$4 + 6$.

559. $4 - 3(-2) = \_\_\_\_\_\_$.

15

$3 - 2(-6)$,
$3 + 12$.

560. Evaluate $3 - 2y$ for $y = -6$.

8

$5 - (-3)$,
$5 + 3$.

561. Evaluate $5 - x$ for $x = -3$.

-3

562. If x is replaced by -3 in the expression $4x^2 + 1$, the result is $4(\_\_\_)^2 + 1$.

37

563. $4(-3)^2 + 1 = 4(9) + 1 = \_\_\_\_\_\_$.

18

$3(-2)^2 + 6$,
$3(4) + 6$,
$12 + 6$.

564. Evaluate $3x^2 + 6$ for $x = -2$.

28

$4 + 3(2)^3$,
$4 + 3(8)$,
$4 + 24$.

565. Evaluate $4 + 3x^3$ for $x = 2$.

-20

$4 + 3(-2)^3$,
$4 + 3(-8)$,
$4 - 24$.

566. Evaluate $4 + 3x^3$ for $x = -2$.

-34

567. If $y = 3$, then $2 - 4y^2 =$_______.

-34

568. If $y = -3$, then $2 - 4y^2 =$_______.

2

569. If $x = 2$, then $6 - x^2 =$_____.

2

$6 - (-2)^2$,
$6 - 4$.

570. If $x = -2$, then $6 - x^2 =$___.

-5; -5

571. When x is replaced by -5 in $3x^2 - x + 1$, the expression becomes $3(\_\_\_)^2 - (\_\_\_) + 1$.

81

572. $3(-5)^2 - (-5) + 1 = 3(25) + 5 + 1 =$______.

18

$2(3)^2 + (3) - 3$,
$2(9) + (3) - 3$,
$18 + 3 - 3$.

573. Evaluate $2x^2 + x - 3$ for $x = 3$.

9

574. Evaluate $3x^2 - 2x + 1$ for $x = 2$.

52

575. Evaluate $5x^2 - 3x - 2$ for $x = -3$.

−30

576. Find the value of $-x^2 - 3x - 2$ if x equals 4.

−6

577. Find the value of $-x^2 - 3x - 2$ if x equals −4.

Remark. Thus far in our evaluations we have used only the operations of multiplication and addition. The next sequence of frames brings division into the picture.

−8; −20

578. If x is replaced by −8 in the expression $\frac{5x}{2}$, the result is $\frac{5(\_\_\_\_)}{2}$ or________.

−4

579. Evaluate $\frac{y}{-3}$ for $y = 12$.

6

580. Evaluate $\frac{3x}{4}$ for $x = 8$.

18

581. Evaluate $\frac{6a}{-3}$ for $a = -9$.

4

582. When x is replaced by 4 in $\frac{2x+1}{3}$, the expression becomes $\frac{2(\_\_\_) + 1}{3}$.

3

583. $\frac{2(4)+1}{3} = \frac{8+1}{3} = \frac{9}{3} =$_____.

-3

$\frac{3(-4)+6}{2}$,

$\frac{-12+6}{2}$,

$\frac{-6}{2}$.

584. If $x = -4$, then $\frac{3x+6}{2} =$ ______.

-2

585. If $x = 3$, then $\frac{2-4x}{5} =$ ______.

5

$\frac{9-(-1)}{2}$,

$\frac{9+1}{2}$,

$\frac{10}{2}$.

586. If $x = -1$, then $\frac{9-x}{2} =$ _____.

8

587. If x is replaced by 8 in the expression $\frac{3x}{4} + 1$, the result is $\frac{3(\_\_\_)}{4} + 1$.

7

588. $\frac{3(8)}{4} + 1 = \frac{24}{4} + 1 = 6 + 1 =$ ___.

10

589. Evaluate $\frac{4y}{3} - 2$ for $y = 9$.

12

590. Evaluate $6 - \frac{3x}{2}$ for $x = -4$.

5

591. Evaluate $\frac{x^2 + 2x}{3}$ for $x = 3$.

0

592. Evaluate $\frac{2x^2 - x}{2}$ for $x = 0$.

3

593. Find the value of $\frac{y^2 + 2}{y}$ if y equals 2.

-14

594. Find the value of $\frac{3y^2 + 8}{y}$ if y equals -4.

-2

595. Find the value of $\frac{x^2 + x}{2} - 3$ if x equals -2.*

Remark. Your next task is to learn to evaluate expressions containing more than one variable.

-4

596. If x is replaced by 3 and y is replaced by -4 in the expression $x + 2y$, the result is $(3) + 2(\_\_\_\_)$.

-5

597. $(3) + 2(-4) =$ _______.

5

$3(2) + (-1)$,
$6 - 1$.

598. Evaluate $3x + y$ for $x = 2$ and $y = -1$.

7

599. Evaluate $x^2 - y$ for $x = -3$ and $y = 2$.

*See Exercise 8, page 85, for additional practice.

8

600. Evaluate $3x^2 + 2y$ for $x = 2$ and $y = -2$.

10

601. If $x = 1$ and $y = -3$, then $x^2 + y^2 =$ ______.

4

602. If $x = -1$ and $y = -1$, then $6x^2 - 2y^2 =$ _____.

22

603. If $x = 2$ and $y = -2$, then $5x^2 - y =$ ______.

-1

604. Find the value of $\frac{x^2 - y^2}{4}$ if $x = 0$ and $y = 2$.

-1

605. Find the value of $\frac{x^2 - y^2}{4}$ if $x = 0$ and $y = -2$.

1; -2; 3

606. When evaluating $2x^2 + y^2 - z^2$ for $x = 1$, $y = -2$, and $z = 3$, the expression becomes
2(___)^2 + (___)^2 − (___)^2.

-3

607. $2(1)^2 + (-2)^2 - (3)^2 = 2 + 4 - 9 =$ ______.

18

608. Evaluate $x^2 + 2y^2 + z$ for $x = 3$, $y = -2$, $z = 1$.

-3

609. Evaluate $2x^2 - y + z$ for $x = 0$, $y = 2$, $z = -1$.

3

610. Evaluate $x - y^2 + z^2$ for $x = -1$, $y = 0$, $z = 2$.

29

611. Evaluate $x + 3y^2 - z^2$ for $x = 2$, $y = -3$, $z = 0$.

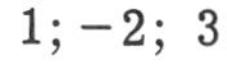

1; −2; 3

612. When evaluating $3xy^2z$ for $x = 1$, $y = -2$, $z = 3$, the expression becomes $3(\_\_\_)(\_\_\_)^2(\_\_\_)$.

36

613. $3(1)(-2)^2(3) = 3(1)(4)(3) =$ _____.

36

614. Evaluate xyz^2 for $x = 2$, $y = 2$, $z = -3$.

−8

615. Evaluate $2x^2yz$ for $x = 1$, $y = -2$, $z = 2$.

−36

616. Evaluate $-2xy^2z^2$ for $x = 2$, $y = 3$, $z = -1$.*

Remark. Numerical evaluation is used in conjunction with formulas to solve many problems of a practical nature. The last few frames of this unit introduce formulas.

formula

617. Relationships between physical quantities are often expressed by formulas. For example, the fact that the area of a rectangle in square units is equal to the product of its length and its width in linear units is expressed by means of the equation $A = lw$. This equation is an example of a ______________.

48

618. To find the area of any rectangle, the numerical values for length (l) and width (w) can be substituted for the respective variables in the formula $A = lw$. For example, if $l = 8$ inches and $w = 6$ inches, $A = (8)(6)$ or ______ square inches.

(3.14)(6)

Since π is only approximately equal to 3.14, this result will only be an approximation.

619. The length of the circumference of a circle is given by $C = \pi d$, where d is the length of the diameter and π is a number that is approximated by 3.14. To evaluate the formula for C if d is 6 inches, 6 is substituted for d and 3.14 for π to give $C = (\_\_\_\_\_)(\_\_\_\_\_)$.

* See Exercise 9, page 85, for additional practice.

18.84

Since the diameter is given in inches, the circumference is also in inches.

620. $C = (3.14)(6) =$ ________.

31.4

621. Evaluate $C = \pi d$ for C if $d = 10$ and 3.14 is used for π.

28.26

622. Evaluate $A = \pi r^2$ for A if $r = 3$ and 3.14 is used for π.

18

623. Evaluate $P = 2(l + w)$ for P if $l = 6$ and $w = 3$.

30

624. Evaluate $A = \frac{1}{2}hb$ for A if $h = 10$ and $b = 6$.

24

625. Evaluate $A = \frac{1}{2}h(a + b)$ for A if $h = 6$, $a = 3$, and $b = 5$.

Remark. A great deal of the work of technicians and engineers involves the evaluation of formulas and, if you continue your study of mathematics, you will find that you will have much use for this skill.

The remainder of this unit is a review, and this last sequence of frames will give you an opportunity to obtain an overview of what you have covered. Although you should not anticipate learning anything you have not learned to this point, the review will highlight the main ideas in a brief way and will reinforce what you have learned.

left

626. A line graph can be used to visualize the relative order of numbers. Of any two numbers, the graph of the smaller always lies to the left of the larger. The graph of -8 will lie to the (left/right) of the graph of -5.

greater than

627. The graphs of -7 and -1 are shown on the line graph.

-1 is (greater than/less than) -7.

positive

628. The absolute value of any number can be associated with the distance from the origin to the graph of the number, consequently the absolute value of any number other than zero must be (positive/negative).

7

629. $|-7| =$ _____.

negative

630. The sum of two positive numbers is a positive number, and the sum of two negative numbers is a ________________ number.

12

631. $(+5) + (+7) =$ _____.

-12

632. $(-5) + (-7) =$ _____.

positive; negative

633. The sum of a positive number and a negative number can be either ____________ or ____________.

2

634. $(+7) + (-5) =$ _____.

-2

635. $(-7) + (+5) =$ _____.

$(-b)$

636. $a - b$ is equal to $a + ($ ____ $)$.

-3

637. $9 - 12$ can be looked at as the difference $(9) - (12)$, or as the sum $(9) + (-12)$. In either case, the expression can be represented by the single numeral______.

$(2) + (5)$

638. $2 - (-5)$ can be looked at as the difference $(2) - (-5)$ or as the sum ______ + ______.

10

639. $4 - (2) - (-3) + 5$ can be represented by the single numeral______.

$9y$

640. By the distributive law, $ab + ac = (b + c)a$. Therefore, expressions containing like terms can be simplified by combining the coefficients of the like terms. For example, $2x - 6x = -4x$, and $4y - 2y + 7y =$______.

$2x^2 + x + x^3 - 3$

641. If an expression involving only addition or subtraction contains a set of parentheses preceded by a positive sign, the expression can be rewritten without the parentheses without further change to the terms in the expression. Thus, $(2x^2 + x) + (x^3 - 3)$ can be rewritten __________________.

$3x^2 + x - x^3 + 2$

642. If an expression involving only addition or subtraction contains a set of parentheses preceded by a negative sign, the expression can be rewritten without the parentheses, if each term within the parentheses is replaced with its negative. Thus, $3x^2 + x - (x^3 - 2) =$__________________.

$21; -72$

643. The product of two integers of like sign is a positive number, and the product of two integers of unlike sign is a negative number. Thus, $(-3)(-7) =$______ and $(-8)(9) =$______.

0

644. The product of 0 and any integer is 0. Thus, $(8)(7)(0) =$ ______.

$-2x^2$; $5x^4$

645. The quotient of any two expressions of like signs carries a positive sign, and the quotient of any two expressions of unlike signs carries a negative sign. $\frac{-24x^2y}{12y} =$ ________ and $\frac{5x^5z}{xz} =$ _______.

0

646. The quotient $\frac{0}{3} = 0$, because $3(0) = 0$. Similarly, $\frac{0}{7} =$ ______.

undefined

647. The quotient $\frac{3}{0}$ is undefined, because there is no number that will multiply by 0 to give 3. Similarly, $\frac{7}{0}$ is ____________________.

undefined

648. $\frac{0}{0}$ is ____________________.

38

45 − 7 = 38.

649. In simplifying any numerical expression, multiplication and division are performed before addition and subtraction. Therefore, $5 \cdot 9 - 7 =$ ______.

−1

650. Grouping devices, such as parentheses or fraction bars, sometimes alter the order of performing operations. For example, in simplifying $\frac{8+4}{3} + 1$, the operation of addition would first be performed on $8 + 4$ to obtain $\frac{12}{3} + 1$. 12 is then divided by 3, and the result added to 1. $\frac{8+4}{3} + 1$, therefore, is equal to 5. Simplify $\frac{9+7}{8} - 3$.

-18

651. In any expression involving powers, the power is usually computed first. Thus
$\frac{3-6^2}{11} = \frac{3-36}{11} = \frac{-33}{11} = -3$, where 6^2 is first written as as 36. Simplify $\frac{5-9^2}{4} + 1$.

-10

652. $\frac{6-4(3)^2}{5-2} =$ ______.

-29

653. The process of finding a numerical value for an algebraic expression when the variables have been replaced by specified numbers is called numerical evaluation. Evaluate $3 - 2y^2$ for $y = 4$.

-11

654. Find the value of $-x^2 + 4x + 1$ if x equals -2.

2

655. Evaluate $\frac{x^2 - 2x + 3}{x}$ for $x = 3$.

2

656. Evaluate $3x^2 - 2y + z^2$ for $x = 0$, $y = 1$, $z = -2$.

Remark. This concludes Unit II. To see what you have learned, you can take one of the self-evaluation tests on pages 87-88. If you completed one form before starting this unit, we suggest that you use the alternate form now.

Exercise 1. If you have difficulty with this exercise, reenter the program at Frame 80.

Simplify.

1. $2 + (-3)$ **2.** $(-4) + (-7)$ **3.** $(-6) + 8$ **4.** $4 + (-5)$ **5.** $4 - 2 + 6$ **6.** $3 - 3 - 2$

7. $-8 + 3 + 1$ **8.** $-7 + 6 - 3$ **9.** $-8 - 1 - 1$ **10.** $-11 -5 + 20$ **11.** $6 - 3 - 3$

12. $8 - 7 + 8$ **13.** $15 - 11 + 3$ **14.** $23 + 1 - 15$ **15.** $-25 + 10 + 16$

16. $-32 + 20 - 2$ **17.** $-7 + 3 + 6 - 2$ **18.** $-8 + 1 - 2 + 12$ **19.** $23 - 15 + 11 - 20$

20. $18 - 15 - 3 + 1$ **21.** $4 - 1 - 9 - 17$ **22.** $-2 + 8 + 12 - 11$ **23.** $16 - 8 - 5 + 12$

24. $14 + 12 - 2 - 18$

Answers

1. -1 **2.** -11 **3.** 2 **4.** -1 **5.** 8 **6.** -2 **7.** -4 **8.** -4 **9.** -10

10. 4 **11.** 0 **12.** 9 **13.** 7 **14.** 9 **15.** 1 **16.** -14 **17.** 0 **18.** 3

19. -1 **20.** 1 **21.** -23 **22.** 7 **23.** 15 **24.** 6

Exercise 2. If you have difficulty with this exercise, reenter the program at Frame 158.

Simplify.

1. $(8) - (5)$ **2.** $(7) - (3)$ **3.** $3 - (-2)$ **4.** $4 - (-4)$ **5.** $(-3) - (2)$

6. $(-3) - (5)$ **7.** $(-6) - (-8)$ **8.** $(-5) - (-5)$ **9.** $2 - 6 - (-2)$ **10.** $3 - 1 - (-4)$

11. $-7 - (-2) - 1$ **12.** $-6 - (-4) - (-2)$ **13.** $8 - (-2) + 4$ **14.** $6 - (-3) + 5$

15. $6 + (-3) - (-2)$

Answers

1. 3 **2.** 4 **3.** 5 **4.** 8 **5.** -5 **6.** -8 **7.** 2 **8.** 0 **9.** -2

10. 6 **11.** -6 **12.** 0 **13.** 14 **14.** 14 **15.** 5

Exercise 3. If you have difficulty with this exercise, reenter the program at Frame 213.

Simplify.

1. $3x + 5x$ **2.** $2y + 6y$ **3.** $4n^2 - 2n^2$ **4.** $-3n^2 + 6n^3$ **5.** $3a + b + a$

6. $2c + 3d + 5c$ **7.** $2x + y - 3x - y$ **8.** $3a - 2a + b - a$ **9.** $x^2 - 3x + 4x - 2x^2$

10. $2z^2 + 3z - z^2 - 3z$ **11.** $-x^2 + 4 + 4x^2 - 1$ **12.** $-2a^3 + 3a + a + a^3$

13. $(3x + y) + (x - 2y)$ **14.** $(4c - 3d) + (-c + 2d)$ **15.** $(-3t + s) + (3t - s)$

16. $(2y - 4z) + (y + 2z)$ **17.** $(x^2 - 2x + 3) + (2x^2 + 2x - 5)$

18. $(z^2 + 3z - 5) + (-z^2 + z + 2)$ **19.** $(-4a^2 + 2a - 1) + (2a^2 - 3a + 2)$

20. $(4n^3 - 2n^2 + 3) + (n^3 + 2n^2 - 4)$

Answers

1. $8x$ **2.** $8y$ **3.** $2n^2$ **4.** $3n^3$ **5.** $4a + b$ **6.** $7c + 3d$ **7.** $-x$ **8.** b

9. $-x^2 + x$ **10.** z^2 **11.** $3x^2 + 3$ **12.** $-a^3 + 4a$ **13.** $4x - y$ **14.** $3c - d$

15. 0 **16.** $3y - 2z$ **17.** $3x^2 - 2$ **18.** $4z - 3$ **19.** $-2a^2 - a + 1$ **20.** $5n^3 - 1$

Exercise 4. If you have difficulty with this exercise, reenter the program at Frame 261.

Simplify.

1. $2x - (x + 1)$ **2.** $3y - (2y + 2)$ **3.** $4a - (2a - 3)$ **4.** $b - (2b + 1)$

5. $(x + 1) - (x - 1)$ **6.** $(z + 3) - (z + 2)$ **7.** $x^2 - (x + 2)$ **8.** $y^2 - (y - 3)$

9. $(x^2 + 2) - (x - 1)$ **10.** $(z^2 - 3) - (z - 2)$ **11.** $(x^2 - 3x) - (x - 1)$

12. $(n^2 + 5n) - (n^2 + n)$ **13.** $(x - 2y) - (x + y) - (x - y)$

14. $(c + 2d) - (2c - d) - (-c + 3d)$ **15.** $(x + 2y - z) - (2y + z)$ **16.** $(a + 2b + c) - (a + b)$

17. $(x^2 + 2x - 3) - (x^2 + x - 4)$ **18.** $(z^2 - 2z + 3) - (2z^2 + z - 5)$

19. $(3a - b + c) - (a - 2b + c)$ **20.** $(4m - n - 3p) - (-m + n - p)$

Answers

1. $x - 1$ **2.** $y - 2$ **3.** $2a + 3$ **4.** $-b - 1$ **5.** 2 **6.** 1 **7.** $x^2 - x - 2$

8. $y^2 - y + 3$ **9.** $x^2 - x + 3$ **10.** $z^2 - z - 1$ **11.** $x^2 - 4x + 1$ **12.** $4n$

13. $-x - 2y$ **14.** 0 **15.** $x - 2z$ **16.** $b + c$ **17.** $x + 1$ **18.** $-z^2 - 3z + 8$

19. $2a + b$ **20.** $5m - 2n - 2p$

Exercise 5. If you have difficulty with this exercise, reenter the program at Frame 289.

Multiply.

1. $(-3)(5)$ **2.** $(2)(-9)$ **3.** $(-6)(-8)$ **4.** $(-5)(-11)$ **5.** $(3)(-7)(1)$

6. $(-3)(-2)(-1)$ **7.** $(4)(-3)(-3)$ **8.** $(-1)(-1)(-5)$ **9.** $(-1)(-1)(-1)(-1)$

10. $(-1)(1)(-1)0$ **11.** $(-8)(2)(-1)(3)$ **12.** $(7)(-2)(-1)(-2)$ **13.** $x^2 \cdot x^3$

14. $y \cdot y^4$ **15.** $z \cdot z^2 \cdot z^3$ **16.** $a^5 \cdot a^7 \cdot a$ **17.** $(2a)(4b)$ **18.** $(3m)(7n)$

19. $(2y^2)(-3z)$ **20.** $(-5r^3)(-2t^2)$ **21.** $(2x)(-3x^2)(xy)$ **22.** $(4a)(a^2b)(-2b^3)$

23. $(-2xy)(3x^2y)(xy^3)$ **24.** $(-x^2z)(2xz)(z^4)$

Answers

1. -15 **2.** -18 **3.** 48 **4.** 55 **5.** -21 **6.** -6 **7.** 36 **8.** -5 **9.** 1

10. 0 **11.** 48 **12.** -28 **13.** x^5 **14.** y^5 **15.** z^6 **16.** a^{13} **17.** $8ab$

18. $21mn$ **19.** $-6y^2z$ **20.** $10r^3t^2$ **21.** $-6x^4y$ **22.** $-8a^3b^4$ **23.** $-6x^4y^5$

24. $-2x^3z^6$

Exercise 6. If you have difficulty with this exercise, reenter the program at Frame 396.

Simplify.

1. $\frac{14}{-2}$ 2. $\frac{-10}{5}$ 3. $\frac{-15}{-3}$ 4. $\frac{0}{-3}$ 5. $\frac{-6}{-6}$ 6. $\frac{24}{-3}$ 7. $\frac{-36}{36}$ 8. $\frac{-8}{0}$

9. $\frac{-16}{-8}$ 10. $\frac{0}{2}$ 11. $\frac{24}{-12}$ 12. $\frac{-30}{-2}$ 13. $\frac{x^3}{x^2}$ 14. $\frac{y^5}{y^2}$ 15. $\frac{4a}{-2}$ 16. $\frac{-6b}{-3}$

17. $\frac{8x^2y}{4xy}$ 18. $\frac{18c^2d^2}{-2c^2d}$ 19. $\frac{-30rs^3}{-5rs^2}$ 20. $\frac{-3m^2n}{3m^2n}$ 21. $\frac{-7s^2t}{-7s^2t}$ 22. $\frac{28rst}{7rs}$

23. $\frac{-35m^2p}{-7mp}$ 24. $\frac{-20a^3b^2c}{4a^2bc}$ 25. $\frac{14x^3y^2z^3}{-7x^2z^3}$ 26. $\frac{0}{3x^2y}$ 27. $\frac{7x^2yz}{0}$ 28. $\frac{-4m^2n^2t}{-4m^2nt}$

Answers

1. -7 **2.** -2 **3.** 5 **4.** 0 **5.** 1 **6.** -8 **7.** -1 **8.** undefined **9.** 2

10. 0 **11.** -2 **12.** 15 **13.** x **14.** y^3 **15.** $-2a$ **16.** $2b$ **17.** $2x$

18. $-9d$ **19.** $6s$ **20.** -1 **21.** 1 **22.** $4t$ **23.** $5m$ **24.** $-5ab$ **25.** $-2xy^2$

26. 0 **27.** undefined **28.** n

Exercise 7. If you have difficulty with this exercise, reenter the program at Frame 475.

Simplify.

1. $3(4) - 10$ 2. $2(3) - 7$ 3. $3(5 - 1) + 3.2$ 4. $2 + (4 + 1) \cdot 2$ 5. $\frac{-6}{2} + (-3)$

6. $\frac{12}{-3} + \frac{-30}{-10}$ 7. $\frac{8+7}{6-1}$ 8. $\frac{2 \cdot 3 + 4 \cdot 4}{10 + 1}$ 9. $7^2 + 2^2$ 10. $(-2)^2 - 3^2$ 11. $\frac{3^2 + 4^2}{7 - 2}$

12. $\frac{4^2 - 5^2}{3^2}$ 13. $\frac{3(2^2) - 4^2}{(-2)^2}$ 14. $\frac{4^2 + 8}{3} - 6$ 15. $[8 - 2(3)^2] \div 5$

16. $[6^2 - 3^2] \div (-9)$

Answers

1. 2 **2.** -1 **3.** 18 **4.** 12 **5.** -6 **6.** -1 **7.** 3 **8.** 2 **9.** 53

10. -5 **11.** 5 **12.** -1 **13.** -1 **14.** 2 **15.** -2 **16.** -3

Exercise 8. If you have difficulty with this exercise, reenter the program at Frame 549.

Evaluate

1. $3x - 7$, for $x = 3$ **2.** $2y + 8$, for $y = -6$ **3.** $6 - z$, for $z = 2$ **4.** $2x^2 - 5$, for $x = 3$

5. $2 + 3b^2$ for $b = -1$ **6.** $5 - 3c^2$, for $c = 2$ **7.** $4x^2 - 2x + 1$, for $x = -1$

8. $2x^2 + 3x - 4$, for $x = -2$ **9.** $-x^2 + x - 5$, for $x = 3$ **10.** $-z^2 + 2z + 3$, for $z = -2$

11. $\frac{2b}{2} - (-1)$, for $b = -3$ **12.** $6 - \frac{4c}{3}$, for $c = 6$ **13.** $\frac{t^2 - 2t}{3}$, for $t = 5$

14. $\frac{n^2 + 3n}{2}$, for $n = -2$ **15.** $\frac{2r^2 - 3r}{5} + 1$, for $r = 4$ **16.** $\frac{s^2 + s}{10} - 2$, for $s = -5$

Answers

1. 2 **2.** −4 **3.** 4 **4.** 13 **5.** 5 **6.** −7 **7.** 7 **8.** −2 **9.** −11

10. −5 **11.** −2 **12.** −2 **13.** 5 **14.** −1 **15.** 5 **16.** 0

Exercise 9. If you have difficulty with this exercise, reenter the program at Frame 596.

Evaluate.

1. $2x + y$, for $x = 3$ and $y = -2$ **2.** $r^2 - s$, for $r = 2$ and $s = 2$

3. $m^2 + n^2$, for $m = 2$ and $n = 1$ **4.** $4a^2 - 3b$, for $a = -3$ and $b = 2$

5. $5y^2 - 4z^2$, for $y = 2$ and $z = -3$ **6.** $2p^2 + 3q$, for $p = 2$ and $q = 4$

7. $\frac{2c - d}{2}$, for $c = 3$ and $d = 2$ **8.** $\frac{x^2 - 3y}{3}$, for $x = 6$ and $y = -1$

9. $2x^2 + y^2 - 3z$, for $x = 1$, $y = -2$, and $z = 4$

10. $a + 2b^2 - c^2$, for $a = -3$, $b = 1$ and $c = 3$

11. $3r + 2s^2 - t^2$, for $r = 0$, $s = -3$, and $t = 2$

Exercise 9. (Continued)

12. $-m^2 + 2n - p^2$, for $m = -2$, $n = 0$, and $p = 1$ **13.** xy^2, for $x = 3$ and $y = -2$

14. b^2c^3, for $b = 2$ and $c = -1$ **15.** $5rs^2t$, for $r = 2$, and $t = 0$

16. $-3x^2yz$, for $x = 2$, $y = 3$, and $z = -1$

Answers

1. 4 **2.** 2 **3.** 5 **4.** 30 **5.** −16 **6.** 20 **7.** 2 **8.** 13 **9.** −6

10. −10 **11.** 14 **12.** −5 **13.** 12 **14.** −4 **15.** 0 **16.** 36

SELF-EVALUATION TEST, FORM A

1. If x represents a positive number, then $-x$ represents a __________ number.

2. Circle the members of $\{-6.3, -2, 0, 41\}$ that are integers.

3. On the line graph

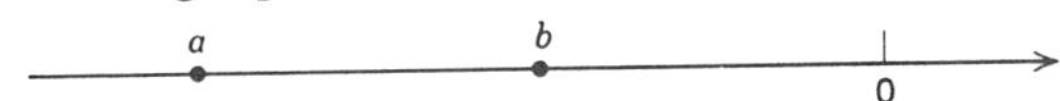

the integer represented by a is (less than/greater than) the integer represented by b.

4. The relative order of the integers 2 and -3 can be shown by writing ____ < ____.

5. $|-3| =$ ____.

6. $6 - 9 + 3 =$ ____.

7. $(4) - (6) =$ ____.

8. $(6) - (-2) =$ ________.

9. $7x^2 - 3x^2 =$ ________.

10. $4x^2 + x^2 - 2x^2 =$ ________.

11. $(-3)(4) =$ ________.

12. $(-2)(-8) =$ ____.

13. $(-5)(0) =$ ____.

14. $(2x)(-x^2)(3x) =$ ________.

15. $\dfrac{6x^3y}{-x^2y} =$ ________.

16. $4 \cdot 2^3 - 6 =$ ____.

17. $\dfrac{3 - (-3)^2}{3} =$ ____.

18. Evaluate $x^2 - 2x + 3$ for $x = 3$.

19. Find the value of $\dfrac{3y^2 + 6}{y}$ if y equals -2.

20. If $x = -2$ and $y = -5$, then $2x^2 - 3y^2 =$ ________.

SELF-EVALUATION TEST, FORM B

1. If y represents a negative number, then $-y$ represents a ____________ number.
2. Circle the members of $\{-7, 4.3, 0, 14\}$ that are integers.
3. On the line graph

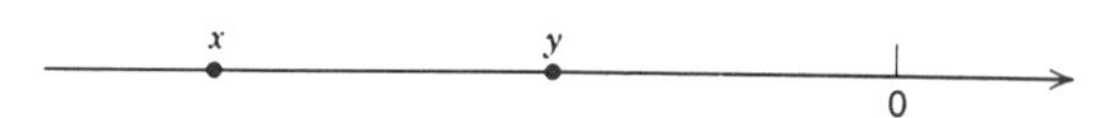

the integer represented by x is (less than/greater than) the integer represented by y

4. The relative order of the integers 6 and -8 can be shown by writing ____ < ____.

Simplify:

5. $|-7| =$ ____.
6. $8 - 10 + 2 =$ ____.
7. $(2) - (8) =$ ____.
8. $(4) - (-3) =$ ____.
9. $3x^2 + 2x^2 - x^2 =$ ________.
10. $(x - 2y) - (2x + y) + (x - y) =$ ________.
11. $(-2)(3) =$ ____.
12. $(-5)(-7) =$ ____.
13. $(0)(-4) =$ ____.
14. $(5x)(-2x^2)(x) =$ ________.
15. $\dfrac{4x^2y}{-xy} =$ ________.
16. $4(3)^2 - 10 =$ ____.
17. $\dfrac{2 - (-2)^2}{2} =$ ____.
18. Evaluate $x^2 - 3x + 1$ for $x = 2$.
19. Find the value of $\dfrac{3x^2 + 6}{x}$ if $x = -3$.
20. If $x = -5$ and $y = -2$, then $3x^2 - 2y^2 =$ ____.

ANSWERS TO TESTS

Form A

1. negative	2. -2; 0; 41	3. less than	4. $-3 < 2$
5. 3	6. 0	7. -2	8. 8
9. $4x^2$	10. $3x^2$	11. -12	12. 16
13. 0	14. $-6x^4$	15. $-6x$	16. 26
17. -2	18. 6	19. -9	20. -67

Form B

1. positive	2. -7; 0; 14	3. less than	4. $-8 < 6$
5. 7	6. 0	7. -6	8. 7
9. $4x^2$	10. $-4y$	11. -6	12. 35
13. 0	14. $-10x^4$	15. $-4x$	16. 26
17. -1	18. -1	19. -11	20. 67

VOCABULARY, UNIT II

The frame in which each word is introduced is shown in parentheses.

absolute value (68)

associative law of addition (108)

coefficient (220)

commutative law of addition (105)

coordinate (6)

difference (158)

distributive law (227)

formula (617)

integers (46)

like terms (226)

line graph (1)

monomial (372)

natural number (8)

negative (12)

negative numbers (16)

numerical coefficient (220)

numerical evaluation (551)

order, of integers (56)

order of operations (475)

origin (5)

positive number (28)

product (216)

quotient (396)

signed number (40)

sum, of integers (82)

term (219)

undefined expression (424)